contents

Acknowledgments...001

Preface..003

Introduction...005

I   The Marginal World.....................................001

II   Patterns of Shore Life 011

III   The Rocky Shores.......................................037

IV   The Rim of Sand...105

V   The Coral Coast 157

VI   The Enduring Sea.......................................201

Appendix: Classification / 205

ACKNOWLEDGMENTS

OUR UNDERSTANDING of the nature of the shore and of the lives of sea animals has been acquired through the labor of many hundreds of people, some of whom have devoted a lifetime to the study of a single group of animals. In my researches for this book I have been deeply conscious of the debt of gratitude we owe these men and women, whose toil allows us to sense the wholeness of life as it is lived by many of the creatures of the shore. I am even more immediately aware of my debt to those I have consulted personally, comparing observations, seeking advice and information and always finding it freely and generously given. It is impossible to express my thanks to all these people by name, but a few must have special mention. Several members of the staff of the United States National Museum have not only settled many of my questions but have given invaluable advice and assistance to Bob Hines in his preparation of the drawings. For this help we are especially grateful to R. Tucker Abbott, Frederick M. Bayer, Fenner Chace, the late Austin H. Clark, Harald Rehder, and Leonard Schultz. Dr. W. N. Bradley of the United States Geological Survey has been my friendly advisor on geological matters, answering many questions and critically reading portions of the manuscript. Professor William Randolph Taylor of the University of Michigan has responded instantly and cheerfully to my calls for aid in identifying marine algae, and Professor and Mrs. T. A. Stephenson of the University College of Wales, whose work on the ecology of the shore has been especially stimulating, have advised and encouraged me in correspondence. To Professor Henry B. Bigelow of Harvard University I am everlastingly in debt for encouragement and friendly counsel over many years. The grant of a Guggenheim Fellowship helped finance the first year of study in which the foundations of this book were laid, and some of the field work that has taken me along the tide lines from Maine to Florida.

PREFACE

LIKE THE SEA ITSELF, the shore fascinates us who return to it, the place of our dim ancestral beginnings. In the recurrent rhythms of tides and surf and in the varied life of the tide lines there is the obvious attraction of movement and change and beauty. There is also, I am convinced, a deeper fascination born of inner meaning and significance.

When we go down to the low-tide line, we enter a world that is as old as the earth itself—the primeval meeting place of the elements of earth and water, a place of compromise and conflict and eternal change. For us as living creatures it has special meaning as an area in or near which some entity that could be distinguished as Life first drifted in shallow waters-reproducing, evolving, yielding that endlessly varied stream of living things that has surged through time and space to occupy the earth.

To understand the shore, it is not enough to catalogue its life. Understanding comes only when, standing on a beach, we can sense the long rhythms of earth and sea that sculptured its land forms and produced the rock and sand of which it is composed; when we can sense with the eye and ear of the mind the surge of life beating always at its shores—blindly, inexorably pressing for a foothold. To understand the life of the shore, it is not enough to pick up an empty shell and say "This is a murex," or "That is an angel wing." True understanding demands intuitive comprehension of the whole life of the creature that once inhabited this empty shell: how it survived amid surf and storms, what were its enemies, how it found food and reproduced its kind, what were its relations to the particular sea world in which it lived.

The seashores of the world may be divided into three basic types: the rugged shores of rock, the sand beaches, and the coral reefs and all their associated features. Each has its typical community of plants and animals. The Atlantic coast of the United States is one of the few in the world that provide clear examples of each of these types. I have chosen it as the setting for my pictures of shore life, although—such is the universality of the sea world—the broad outlines of the pictures might apply-on many shores of the earth.

I have tried to interpret the shore in terms of that essential unity that binds life to the earth. In Chapter I, in a series of recollections of places that have stirred me deeply, I have expressed some of the thoughts and feelings that make the sea's edge, for me, a place of exceeding beauty and fascination. Chapter II introduces as basic themes the sea forces that will recur again and again throughout the book as molding and determining the life of the shore: surf, currents, tides, the very waters of the sea. Chapters III, IV, and V are interpretations, respectively, of a rocky coast, the sand beaches, and the world of the coral reefs.

The drawings by Bob Hines have been provided in abundance so the reader may gain a sense of familiarity with the creatures that move through these pages, and may also be helped to recognize those he meets in his own explorations of the shore. For the convenience of those who like to pigeonhole their findings neatly in the classification schemes the human mind has devised, an appendix presents the conventional groups, or phyla, of plants and animals and describes typical examples. Each form mentioned in the book itself is listed under its Latin as well as its common name in the index.

INTRODUCTION

RACHEL CARSON died in the springtime of 1964, a woman of only fifty-six years, with an established literary reputation and fame, too. She had written four books by that time, all excellent in varying ways and every one a bestseller.

Silent Spring, with its revelations about pesticides and their effects on the natural world, had been the most recent, published less than two years before cancer and its complications took her life. Its popularity with the general reading public—the right book at the right time—made her a pioneer of what we now call environmentalism. This reputation has made a great many people forget that Rachel Carson was first and foremost a writer of considerable literary style whose true love was the sea.

She was, by training, a marine zoologist, and her books before Silent Spring all had been about one aspect or another of the oceans. Part of the reason Silent Spring came to be such a success was that those previous books had established her name. Nevertheless, today her books on the sea seem to be all but forgotten, which is a shame, since in many ways a book such as The Edge of the Sea is more approachable, better written, and more relevant today than the monumental but now somewhat dated Silent Spring.

In October 1955, shortly after The Edge of the Sea was published, John Leonard, already a man who could make words dance, if not yet the gonzo reviewer he was to become, urged "modern city dwellers [who] go down to the sea in bathing suits ...[and] get bored by too much lolling" to buy the book and read it. He wrote that it was "beautifully written, and technically correct." After forty years that is still good advice and not a bad assessment, even though the continuing process of discovery that is science has outdistanced Rachel Carson's facts to some degree.

However, an evaluation in today's intellectual climate would point out that The Edge of the Sea was also a pioneering piece of writing from an ecological perspective, a perspective that was still new and shiny in the 1950s, a perspective that Carson, almost as much as anyone, brought to the reading public's attention as she struggled with her approach to writing this book.

It was a struggle because her original intent had been to write something rather like a field guide, but she soon realized that it was more interesting to write about the relationships among the seashore plants and animals and how the tides and the climate and geological forces affected them.

The book she finally ended up with was, and still is, a pleasure to read. We feel as though a well-informed friend has taken us by the hand as we walk along the ocean's rim and explained all the bits of the world that we see, giving us an understanding of how they fit together and pointing out some other bits that we failed to notice before but always will notice now that we know about them.

Before the turn of the century, the great German zoologist Ernst Haeckel used the term oecology to mean the study of the "economy of animals and plants." It was not until decades into the present century that the study of organisms as part of a community, subject to a changing world—biology in context—gained wide scientific acceptance and entered the biological lexicon as ecology. And it was mid-century before the general public, reading books like Rachel Carson's, began to understand this way of looking at the world as distinct from the older presentation of series of biological life histories, isolated and untouched by external forces.

According to Paul Brooks, the editor of The Edge of the Sea,

Rachel Carson's original plan was to write a series of entries on what is to be found along the sea's edge. The book made from them would have been entitled A Guide to Seashore Life on the Atlantic Coast. It would have been a less integrated, altogether less "ecological" book. But as she began to write, Carson grew more and more uncomfortable with the idea behind the book. The idea had had two parents—a publisher and a writer; in the end the writer got custody of the baby.

The gestation of the idea began when Rosalind Wilson, an editor at Houghton Mifflin, invited a group of literary folks "lacking in biological sophistication" to her home on Cape Cod for the weekend. While walking on the beach they found horseshoe crabs that they believed had been stranded by the storm the night before. They were compassionate, if unknowing, and so they returned all of them to the sea. The horseshoe crabs would have regarded the incident as a terrible setback to their life plan, for they had lumbered ashore to lay their eggs.

When Rosalind Wilson returned to her office in Boston on Monday morning, she sat down and typed out a memo suggesting that Houghton Mifflin find an author who could write a guidebook that "would dispel such ignorance." Soon after, while Rachel Carson was still writing the book that was to be her first bestseller, The Sea Around Us, the proposal for such a guidebook was put to her and she accepted.

The proposal must have sounded to her like a book she had wanted to write for several years. As early as 1948 she had written her literary agent, Marie Rodell, "Among my remote literary projects is a book on the lives of shore animals, which Mr. Teale once asked me to write for his benefit."

In 1950, she wrote to Paul Brooks that for each important form of life the book would include a "biological sketch ... which, while brief, suggests a living creature and illuminates the basic conditions of its life: why it lives where it does, how it has adapted its structures and habitat to its environment, how it gets food, its life cycle, its enemies, competitors, associates." She wanted "to take the seashore out of the category of scenery and make it come alive ... An ecological concept will dominate the book." At Houghton Mifflin, renowned for its excellent field guides, these "biological sketches" must have seemed quite straightforward. But to a writer nothing is straightforward, and to an ecological thinker, which is what Rachel Carson was, the biological sketches developed into something more complicated.

Carson was hard at work on the book when, in 1953, she wrote plaintively to Brooks, "Why is it such agony to put on paper?" Very soon thereafter she wrote to him again. "I decided that I have been trying for a very long time to write the wrong kind of book.... I think we could say that the book has become an interpretation of ... types of shore.... As I am now writing, the routine ... facts, that were so difficult for me to incorporate into the text, are now being saved for the captions ... or for a tabular summary I'd like to tuck in at the end of the book. This solution frees my style to be itself; the attempt to write a structureless chapter that was just one little thumbnail biography after another was driving me mad. I don't know why I once thought I should do it that way, but I did."

Paul Brooks told me that she had been halfway through the writing of the book when she scrapped it and began again to write what became The Edge of the Sea. Lucky she did so; it is a better and more enduring book than A Guide to Seashore Life would have been, and current guidebooks can bring us up to date on recent discoveries to supplement it.

Despite her fame as the author of Silent Spring, Carson's deep interest lay with the ocean, as witnessed not only by her three books concerning it and its shores and her formal education in marine zoology, but also by the fact that as soon as she was able financially to afford it, she bought a property on the coast of Maine—and there she built a home in which she lived for a good part of every year and did much of her writing. At her request, after her death, some of her ashes were scattered off Cape Newagen, near that home.

It wasn't until she was forty-six years old that sales of her second book, The Sea Around Us, allowed her to move to the seashore. As a young woman, still in graduate school at Johns Hopkins, she had begun assuming the financial responsibility for her family, a responsibility that increased over the years as first her mother and then an ailing niece with a son moved in with her. When the niece died, Carson adopted the son. Later she went to work as an aquatic biologist and editor at the U.S. Fish and Wildlife Service and sold free-lance articles wherever she could for whatever fee she could command. It wasn't easy.

She never married.

Rachel Carson was born in 1907 and grew up in rural Springdale, Pennsylvania, slightly northeast of Pittsburgh. Her mother encouraged bookish Rachel in her interests in the natural world. And there she became fascinated with the world's oceans and read what she could find about them. I can testify to a midwesterner's yearning for the sea, for I grew up in that part of the country and to me the ocean came to represent force, power, mystery, and great beauty, a vivid contrast to my everyday world, and I always assumed that someday I would live beside it. It wasn't until I was well into my seventh decade that I acted on that assumption. But now, in a home not too distant from Rachel Carson's, I can watch the tide pull the sea from the shore and return it as I write this introduction.

As a very young woman Carson believed that for a career she would have to choose between her scientific interest in the ocean and her already developed skills in and love of writing. It wasn't until the 1930s that she found a way to blend the two. It was then that she recalled reading Tennyson: "On a night when rain and wind beat against the windows of my college dormitory room, a line from Locksley Hall burned itself into my mind—For the mighty wind arises, roaring seaward, and I go."

Paul Brooks is retired now, but I telephoned him at home one day and asked him if he thought she would have returned to the sea as a subject for additional books had she lived, or whether the success of Silent Spring would have pushed her writing in another direction. "Well I'm not sure," he said. "She'd talked for years about doing a book that was vast and unfocused, that was about Life Itself. I'm glad she never wrote it, because it always sounded too vague and too broad. And although Silent Spring was a tremendous success, she never saw herself as a crusader. She just felt compelled to write that book. But, no, I don't think she was finished with the sea as a subject."

Today we need a Rachel Carson to write about ocean "dead zones," the degradation of ocean habitats, the dying of coral reefs, the effects of global warming on ocean waters. And of the last, the reader will find on the early pages of The Edge of the Sea that Carson was already writing, in the early 1950s, about how ocean life was being changed by warming waters.

Brooks also added that he thought it was significant that the text she asked to be read at her own funeral was one from her writing on the sea, not one from the more recently published Silent Spring. Her request was not honored, but it would have been suitable, for the tone of the passage is elegiac. It begins, "Now I hear the sea sounds about me; the night high tide is rising, swirling with a confused rush of waters against the rocks below my study window...."It comes from the epilogue to The Edge of the Sea, and although these are among the final words of the book, they might be a good place for the reader in 1998 to begin.

Sue Hubbell

Maine

February 1998

Paul Brooks, Rachel Carson's editor and her friend, is the author of The House of Life: Rachel Carson at Work. Both Mr. Brooks's book and his remembrances of Carson were very helpful in writing this introduction. A new edition of his excellent biography is forthcoming from Sierra Club Books. I also consulted Linda Lear, a research professor of environmental history at George Washington University and the foremost authority on Carson's life and work. She is the author of Rachel Carson: Witness for Nature, published in 1997 by Henry Holt and Company.

I
THE MARGINAL WORLD

THE EDGE of the sea is a strange and beautiful place. All through the long history of Earth it has been an area of unrest where waves have broken heavily against the land, where the tides have pressed forward over the continents, receded, and then returned. For no two successive days is the shore line precisely the same. Not only do the tides advance and retreat in their eternal rhythms, but the level of the sea itself is never at rest. It rises or falls as the glaciers melt or grow, as the floor of the deep ocean basins shifts under its increasing load of sediments, or as the earth's crust along the continental margins warps up or down in adjustment to strain and tension. Today a little more land may belong to the sea, tomorrow a little less. Always the edge of the sea remains an elusive and indefinable boundary.

The shore has a dual nature, changing with the swing of the tides, belonging now to the land, now to the sea. On the ebb tide it knows the harsh extremes of the land world, being exposed to heat and cold, to wind, to rain and drying sun. On the flood tide it is a water world, returning briefly to the relative stability of the open sea.

Only the most hardy and adaptable can survive in a region so mutable, yet the area between the tide lines is crowded with plants and animals. In this difficult world of the shore, life displays its enormous toughness and vitality by occupying almost every conceivable niche. Visibly, it carpets the intertidal rocks; or half hidden, it descends into fissures and crevices, or hides under boulders, or lurks in the wet gloom of sea caves. Invisibly, where the casual observer would say there is no life, it lies deep in the sand, in burrows and tubes and passageways. It tunnels into solid rock and bores into peat and clay. It encrusts weeds or drifting spars or the hard, chitinous shell of a lobster. It exists minutely, as the film of bacteria that spreads over a rock surface or a wharf piling; as spheres of protozoa, small as pinpricks, sparkling at the surface of the sea; and as Lilliputian beings swimming through dark pools that lie between the grains of sand.

The shore is an ancient world, for as long as there has been an earth and sea there has been this place of the meeting of land and water. Yet it is a world that keeps alive the sense of continuing creation and of the relentless drive of life. Each time that I enter it, I gain some new awareness of its beauty and its deeper meanings, sensing that intricate fabric of life by which one creature is linked with another, and each with its surroundings.

In my thoughts of the shore, one place stands apart for its revelation of exquisite beauty. It is a pool hidden within a cave that one can visit only rarely and briefly when the lowest of the year's low tides fall below it, and perhaps from that very fact it acquires some of its special beauty. Choosing such a tide, I hoped for a glimpse of the pool. The ebb was to fall early in the morning. I knew that if the wind held from the northwest and no interfering swell ran in from a distant storm the level of the sea should drop below the entrance to the pool. There had been sudden ominous showers in the night, with rain like handfuls of gravel flung on the roof. When I looked out into the early morning the sky was full of a gray dawn light but the sun had not yet risen. Water and air were pallid. Across the bay the moon was a luminous disc in the western sky, suspended above the dim line of distant shore—the full August moon, drawing the tide to the low, low levels of the threshold of the alien sea world. As I watched, a gull flew by, above the spruces. Its breast was rosy with the light of the unrisen sun. The day was, after all, to be fair.

Later, as I stood above the tide near the entrance to the pool, the promise of that rosy light was sustained. From the base of the steep wall of rock on which I stood, a moss-covered ledge jutted seaward into deep water. In the surge at the rim of the ledge the dark fronds of oarweeds swayed, smooth and gleaming as leather. The projecting ledge was the path to the small hidden cave and its pool. Occasionally a swell, stronger than the rest, rolled smoothly over the rim and broke in foam against the cliff. But the intervals between such swells were long enough to admit me to the ledge and long enough for a glimpse of that fairy pool, so seldom and so briefly exposed.

And so I knelt on the wet carpet of sea moss and looked back into the dark cavern that held the pool in a shallow basin. The floor of the cave was only a few inches below the roof, and a mirror had been created in which all that grew on the ceiling was reflected in the still water below.

Under water that was clear as glass the pool was carpeted with green sponge. Gray patches of sea squirts glistened on the ceiling and colonies of soft coral were a pale apricot color. In the moment when I looked into the cave a little elfin starfish hung down, suspended by the merest thread, perhaps by only a single tube foot. It reached down to touch its own reflection, so perfectly delineated that there might have been, not one starfish, but two. The beauty of the reflected images and of the limpid pool itself was the poignant beauty of things that are ephemeral, existing only until the sea should return to fill the little cave.

Whenever I go down into this magical zone of the low water of the spring tides, I look for the most delicately beautiful of all the shore's inhabitants—flowers that are not plant but animal, blooming on the threshold of the deeper sea. In that fairy cave I was not disappointed. Hanging from its roof were the pendent flowers of the hydroid Tubularia, pale pink, fringed and delicate as the wind flower. Here were creatures so exquisitely fashioned that they seemed unreal, their beauty too fragile to exist in a world of crushing force. Yet every detail was functionally useful, every stalk and hydranth and petal-like tentacle fashioned for dealing with the realities of existence. I knew that they were merely waiting, in that moment of the tide's ebbing, for the return of the sea. Then in the rush of water, in the surge of surf and the pressure of the incoming tide, the delicate flower heads would stir with life. They would sway on their slender stalks, and their long tentacles would sweep the returning water, finding in it all that they needed for life.

And so in that enchanted place on the threshold of the sea the realities that possessed my mind were far from those of the land world I had left an hour before. In a different way the same sense of remoteness and of a world apart came to me in a twilight hour on a great beach on the coast of Georgia. I had come down after sunset and walked far out over sands that lay wet and gleaming, to the very edge of the retreating sea. Looking back across that immense flat, crossed by winding, water-filled gullies and here and there holding shallow pools left by the tide, I was filled with awareness that this intertidal area, although abandoned briefly and rhythmically by the sea, is always reclaimed by the rising tide. There at the edge of low water the beach with its reminders of the land seemed far away. The only sounds were those of the wind and the sea and the birds. There was one sound of wind moving over water, and another of water sliding over the sand and tumbling down the faces of its own wave forms. The flats were astir with birds, and the voice of the willet rang insistently. One of them stood at the edge of the water and gave its loud, urgent cry; an answer came from far up the beach and the two birds flew to join each other.

The flats took on a mysterious quality as dusk approached and the last evening light was reflected from the scattered pools and creeks. Then birds became only dark shadows, with no color discernible. Sanderlings scurried across the beach like little ghosts, and here and there the darker forms of the willets stood out. Often I could come very close to them before they would start up in alarm—the sanderlings running, the willets flying up, crying. Black skimmers flew along the ocean's edge silhouetted against the dull, metallic gleam, or they went flitting above the sand like large, dimly seen moths. Sometimes they "skimmed" the winding creeks of tidal water, where little spreading surface ripples marked the presence of small fish.

The shore at night is a different world, in which the very darkness that hides the distractions of daylight brings into sharper focus the elemental realities. Once, exploring the night beach, I surprised a small ghost crab in the searching beam of my torch. He was lying in a pit he had dug just above the surf, as though watching the sea and waiting. The blackness of the night possessed water, air, and beach. It was the darkness of an older world, before Man. There was no sound but the all-enveloping, primeval sounds of wind blowing over water and sand, and of waves crashing on the beach. There was no other visible life— just one small crab near the sea. I have seen hundreds of ghost crabs in other settings, but suddenly I was filled with the odd sensation that for the first time I knew the creature in its own world—that I understood, as never before, the essence of its being. In that moment time was suspended; the world to which I belonged did not exist and I might have been an onlooker from outer space. The little crab alone with the sea became a symbol that stood for life itself—for the delicate, destructible, yet incredibly vital force that somehow holds its place amid the harsh realities of the inorganic world.

The sense of creation comes with memories of a southern coast, where the sea and the mangroves, working together, are building a wilderness of thousands of small islands off the southwestern coast of Florida, separated from each other by a tortuous pattern of bays, lagoons, and narrow waterways. I remember a winter day when the sky was blue and drenched with sunlight; though there was no wind one was conscious of flowing air like cold clear crystal. I had landed on the surf-washed tip of one of those islands, and then worked my way around to the sheltered bay side. There I found the tide far out, exposing the broad mud flat of a cove bordered by the mangroves with their twisted branches, their glossy leaves, and their long prop roots reaching down, grasping and holding the mud, building the land out a little more, then again a little more.

The mud flats were strewn with the shells of that small, exquisitely colored mollusk, the rose tellin, looking like scattered petals of pink roses. There must have been a colony nearby, living buried just under the surface of the mud. At first the only creature visible was a small heron in gray and rusty plumage—a reddish egret that waded across the flat with the stealthy, hesitant movements of its kind. But other land creatures had been there, for a line of fresh tracks wound in and out among the mangrove roots, marking the path of a raccoon feeding on the oysters that gripped the supporting roots with projections from their shells. Soon I found the tracks of a shore bird, probably a sanderling, and followed them a little; then they turned toward the water and were lost, for the tide had erased them and made them as though they had never been.

Looking out over the cove I felt a strong sense of the interchangeability of land and sea in this marginal world of the shore, and of the links between the life of the two. There was also an awareness of the past and of the continuing flow of time, obliterating much that had gone before, as the sea had that morning washed away the tracks of the bird.

The sequence and meaning of the drift of time were quietly summarized in the existence of hundreds of small snails—the mangrove periwinkles—browsing on the branches and roots of the trees. Once their ancestors had been sea dwellers, bound to the salt waters by every tie of their life processes. Little by little over the thousands and millions of years the ties had been broken, the snails had adjusted themselves to life out of water, and now today they were living many feet above the tide to which they only occasionally returned. And perhaps, who could say how many ages hence, there would be in their descendants not even this gesture of remembrance for the sea.

The spiral shells of other snails—these quite minute—left winding tracks on the mud as they moved about in search of food. They were horn shells, and when I saw them I had a nostalgic moment when I wished I might see what Audubon saw, a century and more ago. For such little horn shells were the food of the flamingo, once so numerous on this coast, and when I half closed my eyes I could almost imagine a flock of these magnificent flame birds feeding in that cove, filling it with their color. It was a mere yesterday in the life of the earth that they were there; in nature, time and space are relative matters, perhaps most truly perceived subjectively in occasional flashes of insight, sparked by such a magical hour and place.

There is a common thread that links these scenes and memories—the spectacle of life in all its varied manifestations as it has appeared, evolved, and sometimes died out. Underlying the beauty of the spectacle there is meaning and significance. It is the elusiveness of that meaning that haunts us, that sends us again and again into the natural world where the key to the riddle is hidden. It sends us back to the edge of the sea, where the drama of life played its first scene on earth and perhaps even its prelude; where the forces of evolution are at work today, as they have been since the appearance of what we know as life; and where the spectacle of living creatures faced by the cosmic realities of their world is crystal clear.

II
PATTERNS OF SHORE LIFE

PATTERNS OF LIFE

THE EARLY HISTORY of life as it is written in the rocks is exceedingly dim and fragmentary, and so it is not possible to say when living things first colonized the shore, nor even to indicate the exact time when life arose. The rocks that were laid down as sediments during the first half of the earth's history, in the Archeozoic era, have since been altered chemically and physically by the pressure of many thousands of feet of superimposed layers and by the intense heat of the deep regions to which they have been confined during much of their existence. Only in a few places, as in eastern Canada, are they exposed and accessible for study, but if these pages of the rock history ever contained any clear record of life, it has long since been obliterated.

The following pages—the rocks of the next several hundred million years, known as the Proterozoic era—are almost as disappointing. There are immense deposits of iron, which may possibly have been laid down with the help of certain algae and bacteria. Other deposits—strange globular masses of calcium carbonate—seem to have been formed by lime-secreting algae. Supposed fossils or faint impressions in these ancient rocks have been tentatively identified as sponges, jellyfish, or hard-shelled creatures with jointed legs called arthropods, but the more skeptical or conservative scientists regard these traces as having an inorganic origin.

Suddenly, following the early pages with their sketchy records, a whole section of the history seems to have been destroyed. Sedimentary rocks representing untold millions of years of pre-Cambrian history have disappeared, having been lost by erosion or possibly, through violent changes in the surface of the earth, brought into a location that now is at the bottom of the deep sea. Because of this loss a seemingly unbridgeable gap in the story of life exists.

The scarcity of fossil records in the early rocks and the loss of whole blocks of sediments may be linked with the chemical nature of the early sea and the atmosphere. Some specialists believe that the pre-Cambrian ocean was deficient in calcium or at least in the conditions that make easily possible the secretion of calcium shells and skeletons. If so, its inhabitants must have been for the most part soft-bodied and so not readily fossilized. A large amount of carbon dioxide in the atmosphere and its relative deficiency in the sea would also have affected the weathering of rock, according to geological theory, so that the sedimentary rocks of pre-Cambrian time must have been repeatedly eroded, washed away, and newly sedimented, with consequent destruction of fossils.

When the record is resumed in the rocks of the Cambrian period, which are about half a billion years old, all the major groups of invertebrate animals (including the principal inhabitants of the shore) suddenly appear, fully formed and flourishing. There are sponges and jellyfish, worms of all sorts, a few simple snail-like mollusks, and arthropods. Algae also are abundant, although no higher plants appear. But the basic plan of each of the large groups of animals and plants that now inhabit the shore had been at least projected in those Cambrian seas, and we may suppose, on good evidence, that the strip between the tide lines 500 million years ago bore at least a general resemblance to the intertidal area of the present stage of earth history.

We may suppose also that for at least the preceding halfbillion years those invertebrate groups, so well developed in the Cambrian, had been evolving from simpler forms, although what they looked like we may never know. Possibly the larval stages of some of the species now living may resemble those ancestors whose remains the earth seems to have destroyed or failed to preserve.

During the hundreds of millions of years since the dawn of the Cambrian, sea life has continued to evolve. Subdivisions of the original basic groups have arisen, new species have been created, and many of the early forms have disappeared as evolution has developed others better fitted to meet the demands of their world. A few of the primitive creatures of Cambrian time have representatives today that are little changed from their early ancestors, but these are the exception. The shore, with its difficult and changing conditions, has been a testing ground in which the precise and perfect adaptation to environment is an indispensable condition of survival.

All the life of the shore—the past and the present—by the very fact of its existence there, gives evidence that it has dealt successfully with the realities of its world—the towering physical realities of the sea itself, and the subtle life relationships that bind each living thing to its own community. The patterns of life as created and shaped by these realities intermingle and overlap so that the major design is exceedingly complex.

Whether the bottom of the shallow waters and the intertidal area consists of rocky cliffs and boulders, of broad plains of sand, or of coral reefs and shallows determines the visible pattern of life. A rocky coast, even though it is swept by surf, allows life to exist openly through adaptations for clinging to the firm surfaces provided by the rocks and by other structural provisions for dissipating the force of the waves. The visible evidence of living things is everywhere about—a colorful tapestry of seaweeds, barnacles, mussels, and snails covering the rocks—while more delicate forms find refuge in cracks and crevices or by creeping under boulders. Sand, on the other hand, forms a yielding, shifting substratum of unstable nature, its particles incessantly stirred by the waves, so that few living things can establish or hold a place on its surface or even in its upper layers. All have gone below, and in burrows, tubes, and underground chambers the hidden life of the sands is lived. A coast dominated by coral reefs is necessarily a warm coast, its existence made possible by warm ocean currents establishing the climate in which the coral animals can thrive. The reefs, living or dead, provide a hard surface to which living things may cling. Such a coast is somewhat like one bordered by rocky cliffs, but with differences introduced by smothering layers of chalky sediments. The richly varied tropical fauna of coral coasts has therefore developed special adaptations that set it apart from the life of mineral rock or sand. Because the American Atlantic coast includes examples of all three types of shore, the various patterns of life related to the nature of the coast itself are displayed there with beautiful clarity.

Still other patterns are superimposed on the basic geologic ones. The surf dwellers are different from those who live in quiet waters, even if members of the same species. In a region of strong tides, life exists in successive bands or zones, from the high-water mark to the line of the lowest ebb tides; these zones are obscured where there is little tidal action or on sand beaches where life is driven underground. The currents, modifying temperature and distributing the larval stages of sea creatures, create still another world.

Again the physical facts of the American Atlantic coast are such that the observer of its life has spread before him, almost with the clarity of a well-conceived scientific experiment, a demonstration of the modifying effect of tides, surf, and currents. It happens that the northern rocks, where life is lived openly, lie in the region of some of the strongest tides of the world, those within the area of the Bay of Fundy. Here the zones of life created by the tides have the simple graphic force of a diagram. The tidal zones being obscured on sandy shores, one is free there to observe the effect of the surf. Neither strong tides nor heavy surf visits the southern tip of Florida. Here is a typical coral coast, built by the coral animals and the mangroves that multiply and spread in the calm, warm waters—a world whose inhabitants have drifted there on ocean currents from the West Indies, duplicating the strange tropical fauna of that region.

And over all these patterns there are others created by the sea water itself—bringing or withholding food, carrying substances of powerful chemical nature that, for good or ill, affect the lives of all they touch. Nowhere on the shore is the relation of a creature to its surroundings a matter of a single cause and effect; each living thing is bound to its world by many threads, weaving the intricate design of the fabric of life.

The problem of breaking waves need not be faced by inhabitants of the open ocean, for they can sink into deep water to avoid rough seas. An animal or plant of the shore has no such means of escape. The surf releases all its tremendous energy as it breaks against the shore, sometimes delivering blows of almost incredible violence. Exposed coasts of Great Britain and other eastern Atlantic islands receive some of the most violent surf in the world, created by winds that sweep across the whole expanse of ocean. It sometimes strikes with a force of two tons to the square foot. The American Atlantic coast, being a sheltered shore, receives no such surf, yet even here the waves of winter storms or of summer hurricanes have enormous size and destructive power. The island of Monhegan on the coast of Maine lies unprotected in the path of such storms and receives their waves on its steep seaward-facing cliffs. In a violent storm the spray from breaking waves is thrown over the crest of White Head, about 100 feet above the sea. In some storms the green water of actual waves sweeps over a lower cliff known as Gull Rock. It is about 60 feet high.

The effect of waves is felt on the bottom a considerable distance offshore. Lobster traps set in water nearly 200 feet deep often are shifted about or have stones carried into them. But the critical problem, of course, is the one that exists on or very close to the shore, where waves are breaking. Very few coasts have completely defeated the attempts of living things to gain a foothold. Beaches are apt to be barren if they are composed of loose coarse sand that shifts in the surf and then dries quickly when the tide falls. Others, of firm sand, though they may look barren, actually sustain a rich fauna in their deeper layers. A beach composed of many cobblestones that grind against each other in the surf is an impossible home for most creatures. But the shore formed of rocky cliffs and ledges, unless the surf be of extraordinary force, is host to a large and abundant fauna and flora.

Barnacles are perhaps the best example of successful inhabitants of the surf zone. Limpets do almost as well, and so do the small rock periwinkles. The coarse brown seaweeds called wracks or rockweeds possess species that thrive in moderately heavy surf, while others require a degree of protection. After a little experience one can learn to judge the exposure of any shore merely by identifying its fauna and flora. If, for example, there is a broad area covered by the knotted wrack—a long and slender weed that lies like a tangled mass of cordage when the tide is out—if this predominates, we know the shore is a moderately protected one, seldom visited by heavy surf. If, however, there is little or none of the knotted wrack but instead a zone covered by a rockweed of much shorter stature, branching repeatedly, its fronds flattened and tapering at the ends, then we sense more keenly the presence of the open sea and the crushing power of its surf. For the forked wrack and other members of a community of low-growing seaweeds with strong and elastic tissues are sure indicators of an exposed coast and can thrive in seas the knotted wrack cannot endure. And if, on still another shore, there is little vegetation of any sort, but instead only a rock zone whitened by a living snow of barnacles-thousands upon thousands of them raising their sharp-pointed cones to the smother of the surf—we may be sure this coast is quite unprotected from the force of the sea.

The barnacle has two advantages that allow it to succeed where almost all other life fails to survive. Its low conical shape deflects the force of the waves and sends the water rolling off harmlessly. The whole base of the cone, moreover, is fixed to the rock with natural cement of extraordinary strength; to remove it one has to use a sharp-bladed knife. And so those twin dangers of the surf zone—the threat of being washed away and of being crushed—have little reality for the barnacle. Yet its existence in such a place takes on a touch of the miraculous when we remember this fact: it was not the adult creature, whose shape and firmly cemented base are precise adaptations to the surf, that gained a foothold here; it was the larva. In the turbulence of heavy seas, the delicate larva had to choose its spot on the wave-washed rocks, to settle there, and somehow not be washed away during those critical hours while its tissues were being reorganized in their transformation to the adult form, while the cement was extruded and hardened, and the shell plates grew up about the soft body. To accomplish all this in heavy surf seems to me a far more difficult thing than is required of the spore of a rockweed; yet the fact remains that the barnacles can colonize exposed rocks where the weeds are unable to gain a footing.

The streamlined form has been adopted and even improved upon by other creatures, some of whom have omitted the permanent attachment to the rocks. The limpet is one of these—a simple and primitive snail that wears above its tissues a shell like the hat of a Chinese coolie. From this smoothly sloping cone the surf rolls away harmlessly; indeed, the blows of falling water only press down more firmly the suction cup of fleshy tissue beneath the shell, strengthening its grip on the rock.

Still other creatures, while retaining a smoothly rounded contour, put out anchor lines to hold their places on the rocks. Such a device is used by the mussels, whose numbers in even a limited area may be almost astronomical. The shells of each animal are bound to the rock by a series of tough threads, each of shining silken appearance. The threads are a kind of natural silk, spun by a gland in the foot. These anchor lines extend out in all directions; if some are broken, the others hold while the damaged lines are being replaced. But most of the threads are directed forward and in the pounding of storm surf the mussel tends to swing around and head into the seas, taking them on the narrow "prow" and so minimizing their force.

Even the sea urchins can anchor themselves firmly in moderately strong surf. Their slender tube feet, each equipped with a suction disc at its tip, are thrust out in all directions. I have marveled at the green urchins on a Maine shore, clinging to the exposed rock at low water of spring tides, where the beautiful coralline algae spread a rose-colored crust beneath the shining green of their bodies. At that place the bottom slopes away steeply and when the waves at low tide break on the crest of the slope, they drain back to the sea with a strong rush of water. Yet as each wave recedes, the urchins remain on their accustomed stations, undisturbed.

For the long-stalked kelps that sway in dusky forests just below the level of the spring tides, survival in the surf zone is largely a matter of chemistry. Their tissues contain large amounts of alginic acid and its salts, which create a tensile strength and elasticity able to withstand the pulling and pounding of the waves.

Still others—animal and plant—have been able to invade the surf zone by reducing life to a thin creeping mat of cells. In such form many sponges, ascidians, bryozoans, and algae can endure the force of waves. Once removed from the shaping and conditioning effect of surf, however, the same species may take on entirely different forms. The pale green crumb-of-bread sponge lies flat and almost paper-thin on rocks facing toward the sea; back in one of the deep rock pools its tissues build up into thickened masses, sprinkled with the cone-and-crater structure that is one of the marks of the species. Or the golden-star tunicate may expose a simple sheet of jelly to the waves, though in quiet water it hangs down in pendulous lobes flecked with the starry forms of the creatures that comprise it.

As on the sands almost everything has learned to endure the surf by burrowing down to escape it, so on the rocks some have found safety by boring. Where ancient marl is exposed on the Carolina coast, it is riddled by date mussels. Masses of peat contain the delicately sculptured shells of mollusks called angel wings, seemingly fragile as china, but nevertheless able to bore into clay or rock; concrete piers are drilled by small boring clams; wooden timbers by other clams and isopods. All of these creatures have exchanged their freedom for a sanctuary from the waves, being imprisoned forever within the chambers they have carved.

The vast current systems, which flow through the oceans like rivers, lie for the most part offshore and one might suppose their influence in intertidal matters to be slight. Yet the currents have far-reaching effects, for they transport immense volumes of water over long distances—water that holds its original temperature through thousands of miles of its journey. In this way tropical warmth is carried northward and arctic cold brought far down toward the equator. The currents, probably more than any other single element, are the creators of the marine climate.

The importance of climate lies in the fact that life, even as broadly defined to include all living things of every sort, exists within a relatively narrow range of temperature, roughly between 32° F. and 210° F. The planet Earth is particularly favorable for life because it has a fairly stable temperature. Especially in the sea, temperature changes are moderate and gradual and many animals are so delicately adjusted to the accustomed water climate that an abrupt or drastic change is fatal. Animals living on the shore and exposed to air temperatures at low tide are necessarily a little more hardy, but even these have their preferred range of heat and cold beyond which they seldom stray.

Most tropical animals are more sensitive to change—especially toward higher temperatures—than northern ones, and this is probably because the water in which they live normally varies by only a few degrees throughout the year. Some tropical sea urchins, keyhole limpets, and brittle stars die when the shallow waters heat to about 99° F. The arctic jellyfish Cyanea, on the other hand, is so hardy that it continues to pulsate when half its bell is imprisoned in ice, and may revive even after being solidly frozen for hours. The horseshoe crab is an example of an animal that is very tolerant of temperature change. It has a wide range as a species, and its northern forms can survive being frozen into ice in New England, while its southern representatives thrive in tropical waters of" Florida and southward to Yucatán.

Shore animals for the most part endure the seasonal changes of temperate coasts, but some find it necessary to escape the extreme cold of winter. Ghost crabs and beach fleas are believed to dig very deep holes in the sand and go into hibernation. Mole crabs that feed in the surf much of the year retire to the bottom offshore in winter. Many of the hydroids, so like flowering plants in appearance, shrink down to the very core of their animal beings in winter, withdrawing all living tissues into the basal stalk. Other shore animals, like annuals in the plant kingdom, die at the end of summer. All of the white jellyfish, so common in coastal waters during the summer, are dead when the last autumn gale has blown itself out, but the next generation exists as little plant-like beings attached to the rocks below the tide.

For the great majority of shore inhabitants that continue to live in the accustomed places throughout the year, the most dangerous aspect of winter is not cold but ice. In years when much shore ice is formed, the rocks may be scraped clean of barnacles, mussels, and seaweeds simply by the mechanical action of ice grinding in the surf. After this happens, several growing seasons separated by moderate winters may be needed to restore the full community of living creatures.

Because most sea animals have definite preferences as to aquatic climate, it is possible to divide the coastal waters of eastern North America into zones of life. While variation in the temperature of the water within these zones is in part a matter of the advance from southern to northern latitudes, it is also strongly influenced by the pattern of the ocean currents—the sweep of warm tropical water carried northward in the Gulf Stream, and the chill Labrador Current creeping down from the north on the landward border of the Stream, with complex intermixing of warm and cold water between the boundaries of the currents.

From the point where it pours through the Florida straits up as far as Cape Hatteras, the Stream follows the outer edge of the continental shelf, which varies greatly in width. At Jupiter Inlet on the east coast of Florida this shelf is so narrow that one can stand on shore and look out across emerald-green shallows to the place where the water suddenly takes on the intense blue of the Stream. At about this point there seems to exist a temperature barrier, separating the tropical fauna of southern Florida and the Keys from the warm-temperate fauna of the area lying between Cape Canaveral and Cape Hatteras. Again at Hatteras the shelf becomes narrow, the Stream swings closer inshore, and the northward-moving water filters through a confused pattern of shoals and submerged sandy hills and valleys. Here again is a boundary between life zones, though it is a shifting and far from absolute one. During the winter, temperatures at Hatteras probably forbid the northward passage of migratory warm-water forms, but in summer the temperature barriers break down, the invisible gates open, and these same species may range far toward Cape Cod.

From Hatteras north the shelf broadens, the Stream moves far offshore, and there is a strong infiltration and mixing of colder water from the north, so that the progressive chilling is speeded. The difference in temperature between Hatteras and Cape Cod is as great as one would find on the opposite side of the Atlantic between the Canary Islands and southern Norway—a distance five times as long. For migratory sea fauna this is an intermediate zone, which cold-water forms enter in winter, and warm-water species in summer. Even the resident fauna has a mixed, indeterminate character, for this area seems to receive some of the more temperature-tolerant forms from both north and south, but to have few species that belong to it exclusively.

Cape Cod has long been recognized in zoology as marking the boundary of the range for thousands of creatures. Thrust far into the sea, it interferes with the passage of the warmer waters from the south and holds the cold waters of the north within the long curve of its shore. It is also a point of transition to a different kind of coast. The long sand strands of the south are replaced by rocks, which come more and more to dominate the coastal scene. They form the sea bottom as well as its shores; the same rugged contours that appear in the land forms of this region lie drowned and hidden from view offshore. Here zones of deep water, with accompanying low temperatures, lie generally closer to the shore than they do farther south, with interesting local effects on the populations of shore animals. Despite the deep inshore waters, the numerous islands and the jaggedly indented coast create a large intertidal area and so provide for a rich shore fauna. This is the cold-temperate region, inhabited by many species unable to tolerate the warm water south of the Cape. Partly because of the low temperatures and partly because of the rocky nature of the shore, heavy growths of seaweeds cover the ebb-tide rocks with a blanket of various hues, herds of periwinkles graze, and the shore is here whitened by millions of barnacles or there darkened by millions of mussels.

Beyond, in the waters bathing Labrador, southern Greenland, and parts of Newfoundland, the temperature of the sea and the nature of its flora and fauna are subarctic. Still farther to the north is the arctic province, with limits not yet precisely defined.

Although these basic zones are still convenient and well-founded divisions of the American coast, it became clear by about the third decade of the twentieth century that Cape Cod was not the absolute barrier it had once been for warm-water species attempting to round it from the south. Curious changes have been taking place, with many animals invading this cold-temperate zone from the south and pushing up through Maine and even into Canada. This new distribution is, of course, related to the widespread change of climate that seems to have set in about the beginning of the century and is now well recognized— a general warming-up noticed first in arctic regions, then in subarctic, and now in the temperate areas of northern states. With warmer ocean waters north of Cape Cod, not only the adults but the critically important young stages of various southern animals have been able to survive.

One of the most impressive examples of northward movement is provided by the green crab, once unknown north of the Cape, now familiar to every clam fisherman in Maine because of its habit of preying on the young stages of the clam. Around the turn of the century, zoological manuals gave its range as New Jersey to Cape Cod. In 1905 it was reported near Portland, and by 1930 specimens had been collected in Hancock County, about midway along the Maine coast. During the following decade it moved along to Winter Harbor, and in 1951 was found at Lubec. Then it spread up along the shores of Passamaquoddy Bay and crossed to Nova Scotia.

With higher water temperatures the sea herring is becoming scarce in Maine. The warmer waters may not be the only cause, but they are undoubtedly responsible in part. As the sea herring decline, other kinds of fish are coming in from the south. The menhaden is a larger member of the herring family, used in enormous quantities for manufacturing fertilizer, oils, and other industrial products. In the 1880's there was a fishery for menhaden in Maine, then they disappeared and for many years were confined almost entirely to areas south of New Jersey. About 1950, however, they began to return to Maine waters, followed by Virginia boats and fishermen. Another fish of the same tribe, called the round herring, is also ranging farther north. In the 1920's Professor Henry Bigelow of Harvard University reported it as occurring from the Gulf of Mexico to Cape Cod, and pointed out that it was rare anywhere on the Cape. (Two caught at Provincetown were preserved in the Museum of Comparative Zoology at Harvard.) In the 1950's, however, immense schools of this fish appeared in Maine waters, and the fishing industry began experiments with canning it.

Many other scattered reports follow the same trend. The mantis shrimp, formerly barred by the Cape, has now rounded it and spread into the southern part of the Gulf of Maine. Here and there the soft-shell clam shows signs of being adversely affected by warm summer temperatures and the hard-shell species is replacing it in New York waters. Whiting, once only summer fish north of the Cape, now are caught there throughout the year, and other fish once thought distinctively southern are able to spawn along the coast of New York, where their delicate juvenile stages formerly were killed by the cold winters.

Despite the present exceptions, the Cape Cod—Newfoundland coast is typically a zone of cool waters inhabited by a boreal flora and fauna. It displays strong and fascinating affinities with distant places of the northern world, linked by the unifying force of the sea with arctic waters and with the coasts of the British Isles and Scandinavia. So many of its species are duplicated in the eastern Atlantic that a handbook for the British Isles serves reasonably well for New England, covering probably 80 per cent of the seaweeds and 60 per cent of the marine animals. On the other hand, the American boreal zone has stronger ties with the arctic than does the British coast. One of the large Laminarian seaweeds, the arctic kelp, comes down to the Maine coast but is absent in the eastern Atlantic. An arctic sea anemone occurs in the western North Atlantic abundantly down to Nova Scotia and less numerously in Maine, but on the other side misses Great Britain and is confined to colder waters farther north. The occurrence of many species such as the green sea urchin, the blood-red starfish, the cod, and the herring are examples of a distribution that is circumboreal, extending right around the top of the earth and brought about through the agency of cold currents from melting glaciers and drifting pack ice that carry representatives of the northern faunas down into the North Pacific and North Atlantic.

The existence of so strong a common element between the faunas and floras of the two coasts of the North Atlantic suggests that the means of crossing must be relatively easy. The Gulf Stream carries many migrants away from American shores. The distance to the opposite side is great, however, and the situation is complicated by the short larval life of most species and the fact that shallow waters must be within reach when the time comes for assuming the life of the adult. In this northern part of the Atlantic intermediate way-stations are provided by submerged ridges, shallows, and islands, and the crossing may be broken into easy stages. In some earlier geologic times these shallows were even more extensive, so over long periods both active and involuntary migration across the Atlantic have been feasible.

In lower latitudes the deep basin of the Atlantic must be crossed, where few islands or shallows exist. Even here some transfer of larvae and adults takes place. The Bermuda Islands, after being raised above the sea by volcanic action, received their whole fauna as immigrants from the West Indies via the Gulf Stream. And on a smaller scale the long transatlantic crossings have been accomplished. Considering the difficulties, an impressive number of West Indian species are identical with, or closely related to African species, apparently having crossed in the Equatorial Current. They include species of starfish, shrimp, crayfish, and mollusks. Where such a long crossing has been made it is logical to assume that the migrants were adults, traveling on floating timber or drifting seaweed. In modern times, several African mollusks and starfish have been reported as arriving at the Island of St. Helena by these means.

The records of paleontology provide evidence of the changing shapes of continents and the changing flow of the ocean currents, for these earlier earth patterns account for the otherwise mysterious present distribution of many plants and animals. Once, for example, the West Indian region of the Atlantic was in direct communication, via sea currents, with the distant waters of the Pacific and Indian Oceans. Then a land bridge built up between the Americas, the Equatorial Current turned back on itself to the east, and a barrier to the dispersal of sea creatures was erected. But in species living today we find indications of how it was in the past. Once I discovered a curious little mollusk living in a meadow of turtle grass on the floor of a quiet bay among Florida's Ten Thousand Islands. It was the same bright green as the grass, and its little body was much too large for its thin shell, out of which it bulged. It was one of the scaphanders, and its nearest living relatives are inhabitants of the Indian Ocean. And on the beaches of the Carolinas I have found rocklike masses of calcareous tubes, secreted by colonies of a dark-bodied little worm. It is almost unknown in the Atlantic; again its relatives are Pacific and Indian Ocean forms.

And so transport and wide dispersal are a continuing, universal process—an expression of the need of life to reach out and occupy all habitable parts of the earth. In any age the pattern is set by the shape of the continents and the flow of the currents; but it is never final, never completed.

On a shore where tidal action is strong and the range of the tide is great, one is aware of the ebb and flow of water with a daily, hourly awareness. Each recurrent high tide is a dramatic enactment of the advance of the sea against the continents, pressing up to the very threshold of the land, while the ebbs expose to view a strange and unfamiliar world. Perhaps it is a broad mud flat where curious holes, mounds, or tracks give evidence of a hidden life alien to the land; or perhaps it is a meadow of rockweeds lying prostrate and sodden now that the sea has left them, spreading a protective cloak over all the animal life beneath them. Even more directly the tides address the sense of hearing, speaking a language of their own distinct from the voice of the surf. The sound of a rising tide is heard most clearly on shores removed from the swell of the open ocean. In the stillness of night the strong waveless surge of a rising tide creates a confused tumult of water sounds—swashings and swirlings and a continuous slapping against the rocky rim of the land. Sometimes there are undertones of murmurings and whisperings; then suddenly all lesser sounds are obliterated by a torrential inpouring of water.

On such a shore the tides shape the nature and behavior of life. Their rise and fall give every creature that lives between the high- and low-water lines a twice-daily experience of land life. For those that live near the low-tide line the exposure to sun and air is brief; for those higher on the shore the interval in an alien environment is more prolonged and demands greater powers of endurance. But in all the intertidal area the pulse of life is adjusted to the rhythm of the tides. In a world that belongs alternately to sea and land, marine animals, breathing oxygen dissolved in sea water, must find ways of keeping moist; the few air breathers who have crossed the high-tide line from the land must protect themselves from drowning in the flood tide by bringing with them their own supply of oxygen. When the tide is low there is little or no food for most intertidal animals, and indeed the essential processes of life usually have to be carried on while water covers the shore. The tidal rhythm is therefore reflected in a biological rhythm of alternating activity and quiescence.

On a rising tide, animals that live deep in sand come to the surface, or thrust up the long breathing tubes or siphons, or begin to pump water through their burrows. Animals fixed to rocks open their shells or reach out tentacles to feed. Predators and grazers move about actively. When the water ebbs away the sand dwellers withdraw into the deep wet layers; the rock fauna brings into use all its varied means for avoiding desiccation. Worms that build calcareous tubes draw back into them, sealing the entrance with a modified gill filament that fits like a cork in a bottle. Barnacles close their shells, holding the moisture around their gills. Snails draw back into their shells, closing the doorlike operculum to shut out the air and keep some of the sea's wetness within. Scuds and beach fleas hide under rocks or weeds, waiting for the incoming tide to release them.

All through the lunar month, as the moon waxes and wanes, so the moon-drawn tides increase or decline in strength and the lines of high and low water shift from day to day. After the full moon, and again after the new moon, the forces acting on the sea to produce the tide are stronger than at any other time during the month. This is because the sun and moon then are directly in line with the earth and their attractive forces are added together. For complex astronomical reasons, the greatest tidal effect is exerted over a period of several days immediately after the full and the new moon, rather than at a time precisely coinciding with these lunar phases. During these periods the flood tides rise higher and the ebb tides fall lower than at any other time. These are called the "spring tides" from the Saxon "sprungen." The word refers not to a season, but to the brimming fullness of the water causing it to "spring" in the sense of a strong, active movement. No one who has watched a new-moon tide pressing against a rocky cliff will doubt the appropriateness of the term. In its quarter phases, the moon exerts its attraction at right angles to the pull of the sun so the two forces interfere with each other and the tidal movements are slack. Then the water neither rises as high nor falls as low as on the spring tides. These sluggish tides are called the "neaps"—a word that goes back to old Scandinavian roots meaning "barely touching" or "hardly enough."

On the Atlantic coast of North America the tides move in the so-called semidiurnal rhythm, with two high and two low waters in each tidal day of about 24 hours and 50 minutes. Each low tide follows the previous low by about 12 hours and 25 minutes, although slight local variations are possible. A like interval, of course, separates the high tides.

The range of tide shows enormous differences over the earth as a whole and even on the Atlantic coast of the United States there are important variations. There is a rise and fall of only a foot or two around the Florida Keys. On the long Atlantic coast of Florida the spring tides have a range of 3 to 4 feet, but a little to the north, among the Sea Islands of Georgia, these tides have an 8-foot rise. Then in the Carolinas and northward to New England they move less strongly, with spring tides of 6 feet at Charleston, South Carolina, 3 feet at Beaufort, North Carolina, and 5 feet at Cape May, New Jersey. Nantucket Island has little tide, but on the shores of Cape Cod Bay, less than 30 miles away, the spring tide range is 10 to 11 feet. Most of the rocky coast of New England falls within the zone of the great tides of the Bay of Fundy. From Cape Cod to Passamaquoddy Bay the amplitude of their range varies but is always considerable: 10 feet at Provincetown, 12 at Bar Harbor, 20 at Eastport, 22 at Calais. The conjunction of strong tides and a rocky shore, where much of the life is exposed, creates in this area a beautiful demonstration of the power of the tides over living things.

As day after day these great tides ebb and flow over the rocky rim of New England, their progress across the shore is visibly marked in stripes of color running parallel to the sea's edge. These bands, or zones, are composed of living things and reflect the stages of the tide, for the length of time that a particular level of shore is uncovered determines, in large measure, what can live there. The hardiest species live in the upper zones. Some of the earth's most ancient plants— the blue-green algae—though originating eons ago in the sea, have emerged from it to form dark tracings on the rocks above the high-tide line, a black zone visible on rocky shores in all parts of the world. Below the black zone, snails that are evolving toward a land existence browse on the film of vegetation or hide in seams and crevices in the rocks. But the most conspicuous zone begins at the upper line of the tides. On an open shore with moderately heavy surf, the rocks are whitened by the crowded millions of the barnacles just below the high-tide line. Here and there the white is interrupted by mussels growing in patches of darkest blue. Below them the seaweeds come in— the brown fields of the rockweeds. Toward the low-tide line the Irish moss spreads its low cushioning growth—a wide band of rich color that is not fully exposed by the sluggish movements of some of the neap tides, but appears on all of the greater tides. Sometimes the reddish brown of the moss is splashed with the bright green tangles of another seaweed, a hairlike growth of wiry texture. The lowest of the spring tides reveal still another zone during the last hour of their fall— that sub-tide world where all the rock is painted a deep rose hue by the lime-secreting seaweeds that encrust it, and where the gleaming brown ribbons of the large kelps lie exposed on the rocks.

With only minor variations, this pattern of life exists in all parts of the world. The differences from place to place are related usually to the force of the surf, and one zone may be largely suppressed and another enormously developed. The barnacle zone, for example, spreads its white sheets over all the upper shore where waves are heavy, and the rockweed zone is greatly reduced. With protection from surf, the rockweeds not only occupy the middle shore in profusion but invade the upper rocks and make conditions difficult for the barnacles.

Perhaps in a sense the true intertidal zone is that band between high and low water of the neap tides, an area that is completely covered and uncovered during each tidal cycle, or twice during every day. Its inhabitants are the typical shore animals and plants, requiring some daily contact with the sea but able to endure limited exposure to land conditions.

Above high water of neaps is a band that seems more of earth than of sea. It is inhabited chiefly by pioneering species; already they have gone far along the road toward land life and can endure separation from the sea for many hours or days. One of the barnacles has colonized these higher high-tide rocks, where the sea comes only a few days and nights out of the month, on the spring tides. When the sea returns it brings food and oxygen, and in season carries away the young into the nursery of the surface waters; during these brief periods the barnacle is able to carry on all the processes necessary for life. But it is left again in an alien land world when the last of these highest tides of the fortnight ebbs away; then its only defense is the firm closing of the plates of its shell to hold some of the moisture of the sea about its body. In its life brief and intense activity alternates with long periods of a quiescent state resembling hibernation. Like the plants of the Arctic, which must crowd the making and storing of food, the putting forth of flowers, and the forming of seeds into a few brief weeks of summer, this barnacle has drastically adjusted its way of life so that it may survive in a region of harsh conditions.

Some few sea animals have pushed on even above high water of the spring tides into the splash zone, where the only salty moisture comes from the spray of breaking waves. Among such pioneers are snails of the periwinkle tribe. One of the West Indian species can endure months of separation from the sea. Another, the European rock periwinkle, waits for the waves of the spring tides to cast its eggs into the sea, in almost all activities except the vital one of reproduction being independent of the water.

Below the low water of neaps are the areas exposed only as the rhythmic swing of the tides falls lower and lower, approaching the level of the springs. Of all the intertidal zone this region is linked most closely with the sea. Many of its inhabitants are offshore forms, able to live here only because of the briefness and infrequency of exposure to the air.

The relation between the tides and the zones of life is clear, but in many less obvious ways animals have adjusted their activities to the tidal rhythm. Some seem to be a mechanical matter of utilizing the movement of water. The larval oyster, for example, uses the flow of the tides to carry it into areas favorable for its attachment. Adult oysters live in bays or sounds or river estuaries rather than in water of full oceanic salinity, and so it is to the advantage of the race for the dispersal of the young stages to take place in a direction away from the open sea. When first hatched the larvae drift passively, the tidal currents carrying them now toward the sea, now toward the headwaters of estuaries or bays. In many estuaries the ebb tide runs longer than the flood, having the added push and volume of stream discharge behind it, and the resulting seaward drift over the whole two-week period of larval life would carry the young oysters many miles to sea. A sharp change of behavior sets in, however, as the larvae grow older. They now drop to the bottom while the tide ebbs, avoiding the seaward drift of water, but with the return of the flood they rise into the currents that are pressing upstream, and so are carried into regions of lower salinity that are favorable for their adult life.

Others adjust the rhythm of spawning to protect their young from the danger of being carried into unsuitable waters. One of the tube-building worms living in or near the tidal zone follows a pattern that avoids the strong movements of the spring tides. It releases its larvae into the sea every fortnight on the neap tides, when the water movements are relatively sluggish; the young worms, which have a very brief swimming stage, then have a good chance of remaining within the most favorable zone of the shore.

There are other tidal effects, mysterious and intangible. Sometimes spawning is synchronized with the tides in a way that suggests response to change of pressure or to the difference between still and flowing water. A primitive mollusk called the chiton spawns in Bermuda when the low tide occurs early in the morning, with the return flow of water setting in just after sunrise. As soon as the chitons are covered with water they shed their spawn. One of the Japanese nereid worms spawns only on the strongest tides of the year, near the new- and full-moon tides of October and November, presumably stirred in some obscure way by the amplitude of the water movements.

Many other animals, belonging to quite unrelated groups throughout the whole range of sea life, spawn according to a definitely fixed rhythm that may coincide with the full moon or the new moon or its quarters, but whether the effect is produced by the altered pressure of the tides or the changing light of the moon is by no means clear. For example, there is a sea urchin in Tortugas that spawns on the night of the full moon, and apparently only then. Whatever the stimulus may be, all the individuals of the species respond to it, assuring the simultaneous release of immense numbers of reproductive cells. On the coast of England one of the hydroids, an animal of plant-like appearance that produces tiny medusae or jellyfish, releases these medusae during the moon's third quarter. At Woods Hole on the Massachusetts coast a clamlike mollusk spawns heavily between the full and the new moon but avoids the first quarter. And a nereid worm at Naples gathers in its nuptial swarms during the quarters of the moon but never when the moon is new or full; a related worm at Woods Hole shows no such correlation although exposed to the same moon and to stronger tides.

In none of these examples can we be sure whether the animal is responding to the tides or, as the tides themselves do, to the influence of the moon. With plants, however, the situation is different, and here and there we find scientific confirmation of the ancient and world-wide belief in the effect of moonlight on vegetation. Various bits of evidence suggest that the rapid multiplication of diatoms and other members of the plant plankton is related to the phases of the moon.

Certain algae in river plankton reach the peak of their abundance at the full moon. One of the brown seaweeds on the coast of North Carolina releases its reproductive cells only on the full moon, and similar behavior has been reported for other seaweeds in Japan and other parts of the world. These responses are generally explained as the effect of varying intensities of polarized light on protoplasm.

Other observations suggest some connection between plants and the reproduction and growth of animals. Rapidly maturing herring collect around the edge of concentrations of plant plankton, although the fully adult herring may avoid them. Spawning adults, eggs, and young of various other marine creatures are reported to occur more often in dense phytoplankton than in sparse patches. In significant experiments, a Japanese scientist discovered he could induce oysters to spawn with an extract obtained from sea lettuce. The same seaweed produces a substance that influences growth and multiplication of diatoms, and is itself stimulated by water taken from the vicinity of a heavy growth of rockweeds.

The whole subject of the presence in sea water of the so-called "ectocrines" (external secretions or products of metabolism) has so recently become one of the frontiers of science that actual information is fragmentary and tantalizing. It appears, however, that we may be on the verge of solving some of. the riddles that have plagued men's minds for centuries. Though the subject lies in the misty borderlands of advancing knowledge, almost everything that in the past has been taken for granted, as well as problems considered insoluble, bear renewed thought in the light of the discovery of these substances.

In the sea there are mysterious comings and goings, both in space and time: the movements of migratory species, the strange phenomenon of succession by which, in one and the same area, one species appears in profusion, flourishes for a time, and then dies out, only to have its place taken by another and then another, like actors in a pageant passing before our eyes. And there are other mysteries. The phenomenon of "red tides" has been known from early days, recurring again and again down to the present time—a phenomenon in which the sea becomes discolored because of the extraordinary multiplication of some minute form, often a dinoflagellate, and in which there are disastrous side effects in the shape of mass mortalities among fish and some of the invertebrates. Then there is the problem of curious and seemingly erratic movements of fish, into or away from certain areas, often with sharp economic consequences. When the so-called "Atlantic water" floods the south coast of England, herring become abundant within the range of the Plymouth fisheries, certain characteristic plankton animals occur in profusion, and certain species of invertebrates flourish in the intertidal zone. When, however, this water mass is replaced by Channel water, the cast of characters undergoes many changes.

In the discovery of the biological role played by the sea water and all it contains, we may be about to reach an understanding of these old mysteries. For it is now clear that in the sea nothing lives to itself. The very water is altered, in its chemical nature and in its capacity for influencing life processes, by the fact that certain forms have lived within it and have passed on to it new substances capable of inducing far-reaching effects. So the present is linked with past and future, and each living thing with all that surrounds it.

III

THE ROCKY SHORES

WHEN THE TIDE is high on a rocky shore, when its brimming fullness creeps up almost to the bayberry and the junipers where they come down from the land, one might easily suppose that nothing at all lived in or on or under these waters of the sea's edge. For nothing is visible. Nothing except here and there a little group of herring gulls, for at high tide the gulls rest on ledges of rock, dry above the surf and the spray, and they tuck their yellow bills under their feathers and doze away the hours of the rising water. Then all the creatures of the tidal rocks are hidden from view, but the gulls know what is there, and they know that in time the water will fall away again and give them entrance to the strip between the tide lines.

When the tide is rising the shore is a place of unrest, with the surge leaping high over jutting rocks and running in lacy cascades of foam over the landward side of massive boulders. But on the ebb it is more peaceful, for then the waves do not have behind them the push of the inward pressing tides. There is no particular drama about the turn of the tide, but presently a zone of wetness shows on the gray rock slopes, and offshore the incoming swells begin to swirl and break over hidden ledges. Soon the rocks that the high tide had concealed rise into view and glisten with the wetness left on them by the receding water.

Small, dingy snails move about over rocks that are slippery with the growth of infinitesimal green plants; the snails scraping, scraping, scraping to find food before the surf returns.

Like drifts of old snow no longer white, the barnacles come into view; they blanket rocks and old spars wedged into rock crevices, and their sharp cones are sprinkled over empty mussel shells and lobster-pot buoys and the hard stipes of deep-water seaweeds, all mingled in the flotsam of the tide.

Meadows of brown rockweeds appear on the gently sloping rocks of the shore as the tide imperceptibly ebbs. Smaller patches of green weed, stringy as mermaids' hair, begin to turn white and crinkly where the sun has dried them.

Now the gulls, that lately rested on the higher ledges, pace with grave intentness along the walls of rock, and they probe under the hanging curtains of weed to find crabs and sea urchins.

In the low places little pools and gutters are left where the water trickles and gurgles and cascades in miniature waterfalls, and many of the dark caverns between and under the rocks are floored with still mirrors holding the reflections of delicate creatures that shun the light and avoid the shock of waves—the cream-colored flowers of the small anemones and the pink fingers of soft coral, pendent from the rocky ceiling.

In the calm world of the deeper rock pools, now undisturbed by the tumult of incoming waves, crabs sidle along the walls, their claws busily touching, feeling, exploring for bits of food. The pools are gardens of color composed of the delicate green and ocher-yellow of encrusting sponge, the pale pink of hydroids that stand like clusters of fragile spring flowers, the bronze and electric-blue gleams of the Irish moss, the old-rose beauty of the coralline algae.

And over it all there is the smell of low tide, compounded of the faint, pervasive smell of worms and snails and jellyfish and crabs— the sulphur smell of sponge, the iodine smell of rockweed, and the salt smell of the rime that glitters on the sun-dried rocks.

One of my own favorite approaches to a rocky seacoast is by a rough path through an evergreen forest that has its own peculiar enchantment. It is usually an early morning tide that takes me along that forest path, so that the light is still pale and fog drifts in from the sea beyond. It is almost a ghost forest, for among the living spruce and balsam are many dead trees-some still erect, some sagging earthward, some lying on the floor of the forest. All the trees, the living and the dead, are clothed with green and silver crusts of lichens. Tufts of the bearded lichen or old man's beard hang from the branches like bits of sea mist tangled there. Green woodland mosses and a yielding carpet of reindeer moss cover the ground. In the quiet of that place even the voice of the surf is reduced to a whispered echo and the sounds of the forest are but the ghosts of sound—the faint sighing of evergreen needles in the moving air; the creaks and heavier groans of half-fallen trees resting against their neighbors and rubbing bark against bark; the light rattling fall of a dead branch broken under the feet of a squirrel and sent bouncing and ricocheting earthward.

But finally the path emerges from the dimness of the deeper forest and comes to a place where the sound of surf rises above the forest sounds—the hollow boom of the sea, rhythmic and insistent, striking against the rocks, falling away, rising again.

Up and down the coast the line of the forest is drawn sharp and clean on the edge of a seascape of surf and sky and rocks. The softness of sea fog blurs the contours of the rocks; gray water and gray mists merge offshore in a dim and vaporous world that might be a world of creation, stirring with new life.

The sense of newness is more than illusion born of the early morning light and the fog, for this is in very fact a young coast. It was only yesterday in the life of the earth that the sea came in as the coast subsided, filling the valleys and rising about the slopes of the hills, creating these rugged shores where rocks rise out of the sea and evergreen forests come down to the coastal rocks. Once this shore was like the ancient land to the south, where the nature of the coast has changed little during the millions of years since the sea and the wind and the rain created its sands and shaped them into dune and beach and offshore bar and shoal. The northern coast, too, had its flat coastal plain bordered by wide beaches of sand. Behind these lay a landscape of rocky hills alternating with valleys that had been worn by streams and deepened and sculptured by glaciers. The hills were formed of gneiss and other crystalline rocks resistant to erosion; the lowlands had been created in beds of weaker rocks like sandstones, shale, and marl.

Then the scene changed. From a point somewhere in the vicinity of Long Island the flexible crust of the earth tilted downward under the burden of a vast glacier. The regions we know as eastern Maine and Nova Scotia were pressed down into the earth, some areas being carried as much as 1200 feet beneath the sea. All of the northern coastal plain was drowned. Some of its more elevated parts are now offshore shoals, the fishing banks off the New England and Canadian coasts—Georges, Browns, Quereau, the Grand Bank. None of it remains above the sea except here and there a high and isolated hill, like the present island of Monhegan, which in ancient times must have stood above the coastal plain as a bold monadnock.

Where the mountainous ridges and the valleys lay at an angle to the coast, the sea ran far up between the hills and occupied the valleys. This was the origin of the deeply indented and exceedingly irregular coast that is characteristic of much of Maine. The long narrow estuaries of the Kennebec, the Sheepscot, the Damariscotta and many other rivers run inland a score of miles. These salt-water rivers, now arms of the sea, are the drowned valleys in which grass and trees grew in a geologic yesterday. The rocky, forested ridges between them probably looked much as they do today. Offshore, chains of islands jut out obliquely into the sea, one beyond another—half-submerged ridges of the former land mass.

But where the shore line is parallel to the massive ridges of rock the coast line is smoother, with few indentations. The rains of earlier centuries cut only short valleys into the flanks of the granite hills, and so when the sea rose there were created only a few short, broad bays instead of long winding ones. Such a coast occurs typically in southern Nova Scotia, and also may be seen in the Cape Ann region of Massachusetts, where the belts of resistant rock curve eastward along the coast. On such a coast, islands, where they occur, lie parallel to the shore line instead of putting boldly out to sea.

As geologic events are reckoned, all this happened rather rapidly and suddenly, with no time for gradual adjustment of the landscape;

also it happened quite recently, the present relation of land and sea being achieved perhaps no more than ten thousand years ago. In the chronology of Earth, a few thousand years are as nothing, and in so brief a time the waves have prevailed little against the hard rocks that the great ice sheet scraped clean of loose rock and ancient soil, and so have scarcely marked out the deep notches that in time they will cut in the cliffs.

For the most part, the ruggedness of this coast is the ruggedness of the hills themselves. There are none of the wave-cut stacks and arches that distinguish older coasts or coasts of softer rock. In a few, exceptional places the work of the waves may be seen. The south shore of Mount Desert Island is exposed to heavy pounding by surf; there the waves have cut out Anemone Cave and are working at Thunder Hole to batter through the roof of the small cave into which the surf roars at high tide.

In places the sea washes the foot of a steep cliff produced by the shearing effect of earth pressure along fault lines. Cliffs on Mount Desert—Schooner Head, Great Head, and Otter-tower a hundred feet or more above the sea. Such imposing structures might be taken for wave-cut cliffs if one did not know the geologic history of the region.

On the coasts of Cape Breton Island and New Brunswick the situation is very different and examples of advanced marine erosion occur on every hand. Here the sea is in contact with weak rock lowlands formed in the Carboniferous period. These shores have little resistance to the erosive power of the waves, and the soft sandstone and conglomerate rocks are being cut back at an annual rate averaging five or six inches, or in some places several feet. Marine stacks, caves, chimneys, and archways are common features of these shores.

Here and there on the predominantly rocky coast of northern New England there are small beaches of sand, pebbles, or cobblestones. These have a varied origin. Some came from glacial debris that covered the rocky surface when the land tilted and the sea came in. Boulders and pebbles often are carried in from deeper water offshore by seaweeds that have gripped them firmly with their "holdfasts." Storm waves then dislodge weed and stone and cast them on the shore. Even without the aid of weeds, waves carry in a considerable volume of sand, gravel, shell fragments, and even boulders. These occasional sandy or pebbly beaches are almost always in protected, incurving shores or dead-end coves, where the waves can deposit debris but from which they cannot easily remove it.

When, on those coastal rocks between the serrate line of spruces and the surf, the morning mists conceal the lighthouses and fishing boats and all other reminders of man, they also blur the sense of time and one might easily imagine that the sea came in only yesterday to create this particular line of coast. Yet the creatures that inhabit the intertidal rocks have had time to establish themselves here, replacing the fauna of the beaches of sand and mud that probably bordered the older coast. Out of the same sea that rose over the northern coast of New England, drowning the coastal plain and coming to rest against the hard uplands, the larvae of the rock dwellers came—the blindly searching larvae that drift in the ocean currents ready to colonize whatever suitable land may lie in their path or to die, if no such landfall is their lot.

Although no one recorded the first colonist or traced the succession of living forms, we may make a fairly confident guess as to the pioneers of the occupation of these rocks, and the forms that followed them. The invading sea must have brought the larvae and young of many kinds of shore animals, but only those able to find food could survive on the new shore. And in the beginning the only available food was the plankton that came in renewed clouds with every tide that washed the coastal rocks. The first permanent inhabitants must have been such plankton-strainers as the barnacles and mussels, who require little but a firm place to which they may attach themselves. Around and among the white cones of the barnacles and the dark shells of the mussels it is probable that the spores of algae settled, so that a living green film began to spread over the upper rocks. Then the grazers could come—the little herds of snails that laboriously scrape the rocks with their sharp tongues, licking off the nearly invisible covering of tiny plant cells. Only after the establishment of the plankton-strainers and the grazers could the carnivores settle and survive. The predatory dog whelks, the starfish, and many of the crabs and worms must, then, have been comparative latecomers to this rocky shore. But all of them are there now, living out their lives in the horizontal zones created by the tides, or in the little pockets or communities of life established by the need to take shelter from surf, or to find food, or to hide from enemies.

The pattern of life spread before me when I emerge from that forest path is one characteristic of exposed shores. From the edge of the spruce forests down to the dark groves of the kelps, the life of the land grades into the life of the sea, perhaps with less abruptness than one would expect, for by various little interlacing ties the ancient unity of the two is made clear.

Lichens live in the forest above the sea, in the silent intensity of their toil crumbling away the rocks as lichens have done for millions of years. Some leave the forest and advance over the bare rock toward the tide line; a few go even farther, enduring a periodic submersion by the sea so that they may work their strange magic on the rocks of the intertidal zone. In the dampness of foggy mornings the rock tripe on the seaward slopes is like sheets of thin, pliable green leather, but by midday under a drying sun it has become blackened and brittle; then the rocks look as though they were sloughing off a thin outer layer. Thriving in the salt spray, the wall lichen spreads its orange stain on the cliffs and even on the landward side of boulders that are visited by the highest tides of each moon. Scales of other lichens, sage-green, rolled and twisted into strange shapes, rise from the lower rocks; from their under surfaces black, hairy processes work down among the minute particles of rock substance, giving off an acid secretion to dissolve the rock. As the hairs absorb moisture and swell, fine grains of the rock are dislodged and so the work of creating soil from the rock is advanced.

Below the forest's edge the rock is white or gray or buff, according to its mineral nature. It is dry and belongs to the land; except for a few insects or other land creatures using it as pathways to the sea it is barren. But just above the area that clearly belongs to the sea, it shows a strange discoloration, being strongly marked with streaks or patches or continuous bands of black. Nothing about this black zone suggests life; one would call it a dark stain, or at most a felty roughening of the rock surface. Yet it is actually a dense growth of minute plants. The species that compose it sometimes include a very small lichen, sometimes one or more of the green algae, but most numerously the simplest and most ancient of all plants, the blue-green algae. Some are enclosed in slimy sheaths that protect them from drying and fit them to endure long exposure to sun and air. All are so minute as to be invisible as individual plants. Their gelatinous sheaths and the fact that the whole area receives the spray of breaking waves make this entrance to the sea world slippery as the smoothest ice.

This black zone of the shore has a meaning above and beyond its drab and lifeless aspect—a meaning obscure, elusive, and infinitely tantalizing. Wherever rocks meet the sea, the microplants have written their dark inscription, a message only partially legible although it seems in some way to be concerned with the universality of tides and oceans. Though other elements of the intertidal world come and go, this darkening stain is omnipresent. The rockweeds, the barnacles, the snails, and the mussels appear and disappear in the intertidal zone according to the changing nature of their world, but the black inscriptions of the microplants are always there. Seeing them here on this Maine coast, I remember how they also blackened the coral rim of Key Largo, and streaked the smooth platform of coquina at St. Augustine, and left their tracings on the concrete jetties at Beaufort. It is the same all over the world—from South Africa to Norway and from the Aleutians to Australia. This is the sign of the meeting of land and sea.

Once below the dark film, I begin to look for the first of the sea creatures pressing up to the threshold of the land. In seams and crevices in the high rocks I find them—the smallest of the periwinkle tribe, the rock or rough periwinkle. Some—the infant snails—are so small that I need my hand lens to see them clearly, and among the hundreds that crowd into these cracks and depressions I can find a gradation of sizes up to the half-inch adults. If these were sea creatures of ordinary habits, I would think the small snails were young produced by some distant colony and drifted here as larvae after spending a period at sea. But the rough periwinkle sends no young into the sea; instead it is a viviparous species and the eggs, each encased within a cocoon, are held within the mother while they develop. The contents of the cocoon nourish the young snail until finally it breaks through the egg capsule and then emerges from the mother's body, a completely shelled little creature about the size of a grain of finely ground coffee. So small an animal might easily be washed out to sea; hence, no doubt, the habit of hiding in crevices and in empty barnacle shells, where often I have found them in numbers.

At the level where most of the rough periwinkles live, however, the sea comes only every fortnight on the spring tides, and in the long intervals the flying spray of breaking waves is their only contact with the water. While the rocks are thoroughly wet with spray the periwinkles can spend much time out on the rocks feeding, often working well up into the black zone. The microplants that create the slippery film on the rocks are their food; like all snails of their group, the periwinkles are vegetarians. They feed by scraping the rocks with a peculiar organ set with many rows of sharp, calcareous teeth. This organ, the radula, is a continuous belt or ribbon that lies on the floor of the pharynx. If unwound, it would be many times the length of the animal, but it is tightly coiled like a watch spring. The radula itself consists of chitin, the substance of insects' wings and lobsters' shells. The teeth that stud it are arranged in several hundred rows (in another species, the common periwinkle, the teeth total about 3500). A certain amount of wear is involved in scraping the rocks, and when the teeth in current use are worn down, an endless supply of new ones can be rolled up from behind.

And there is wear, also, on the rocks. Over the decades and the centuries, a large population of periwinkles scraping the rocks for food has a pronounced erosive effect, cutting away rock surfaces, grain by grain, deepening the tide pools. In a tide pool observed for sixteen years by a California biologist, periwinkles lowered the floor about three-eighths of an inch. Rain, frost, and floods—the earth's major forces of erosion-operate on approximately such a scale.

The periwinkles grazing on the intertidal rocks, waiting for the return of the tide, are poised also in time, waiting for the moment when they can complete their present phase of evolution and move forward onto the land. All snails that are now terrestrial came of marine ancestry, their forebears having at some time made the transitional crossing of the shore. The periwinkles now are in mid-passage. In the structure and habits of the three species found on the New England coast, one can see clearly the evolutionary stages by which a marine creature is transformed into a land dweller. The smooth periwinkle, still bound to the sea, can endure only brief exposure. At low tide it remains in wet seaweeds. The common periwinkle often lives where it is submerged only briefly at high tide. It still sheds eggs into the sea and so is not ready for land life. The rough periwinkle, however, has cut most of the ties that confine it to the sea; it is now almost a land animal. By becoming viviparous it has progressed beyond dependence on the sea for reproduction. It is able to thrive at the level of the high water of the spring tides because, unlike the related periwinkles of lower tidal levels, it possesses a gill cavity that is well supplied with blood vessels and functions almost as a lung to breathe oxygen from the air. Constant submersion is, in fact, fatal to it and at the present stage of its evolution it can endure up to thirty-one days of exposure to dry air.

The rough periwinkle has been found by a French experimenter to have the rhythm of the tides deeply impressed upon its behavior patterns, so that it "remembers" even when no longer exposed to the alternating rise and fall of the water. It is most active during the fortnightly visits of the spring tides to its rocks, but in the waterless intervals it becomes progressively more sluggish and its tissues undergo a certain desiccation. With the return of the spring tides the cycle is reversed. When taken into a laboratory the snails for many months reflect in their behavior the advance and retreat of the sea over their native shores.

On this exposed New England coast the most conspicuous animals of the high-tide zone are the rock or acorn barnacles, which are able to live in all but the most tumultuous surf. The rockweeds here are so stunted by wave action that they offer no competition, and so the barnacles have taken over the upper shore, except for such space as the mussels have been able to hold.

At low tide the barnacle-covered rocks seem a mineral landscape carved and sculptured into millions of little sharply pointed cones. There is no movement, no sign or suggestion of life. The stony shells, like those of mollusks, are calcareous and are secreted by the invisible animals within. Each cone-shaped shell consists of six neatly fitted plates forming an encircling ring. A covering door of four plates closes to protect the barnacle from drying when the tide has ebbed, or swings open to allow it to feed. The first ripples of incoming tide bring the petrified fields to life. Then, if one stands ankle-deep in water and observes closely, one sees tiny shadows flickering everywhere over the submerged rocks. Over each individual cone, a feathered plume is regularly thrust out and drawn back within the slightly opened portals of the central door—the rhythmic motions by which the barnacle sweeps in diatoms and other microscopic life of the returning sea.

The creature inside each shell is something like a small pinkish shrimp that lies head downward, firmly cemented to the base of this chamber it cannot leave. Only the appendages are ever exposed—six pairs of branched, slender wands, jointed and set with bristles. Acting together, they form a net of great efficiency.

The barnacle belongs to the group of arthropods known as the Crustacea, a varied horde including the lobsters, crabs, sand hoppers, brine shrimps, and water fleas. The barnacle is different from all related forms, however, in its fixed and sedentary existence. When and how it assumed such a way of life is one of the riddles of zoology, the transitional forms having been lost somewhere in the mists of the past. Some faint suggestions of a similar manner of life—the waiting in a fixed place for the sea to bring food—are found among the amphipods, another group of crustaceans. Some of these spin little webs or cocoons of natural silk and seaweed fibers; though remaining free to come and go they spend much of their time within them, taking their food from the currents. Another amphipod, a Pacific coast species, burrows into colonies of the tunicate called the sea pork, hollowing out for itself a chamber in the tough, translucent substance of its host. Lying in this excavation, it draws currents of sea water over its body and extracts the food.

However the barnacle became what it is, its larval stages clearly proclaim its crustacean ancestry, although early zoologists who looked at its hard shells labeled it a mollusk. The eggs develop inside the parent's shell and presently hatch into the sea in milky clouds of larvae. (The British zoologist Hilary Moore, after studying barnacles on the Isle of Man, estimated a yearly production of a million million larvae from a little over half a mile of shore.) Larval life lasts about three months in the rock barnacle, with several molts and transformations of form. At first the larva, a little swimming creature called a nauplius, is indistinguishable from the larva of all other crustaceans. It is nourished by large globules of fat that not only feed it but keep it near the surface. As the fat globules dwindle, the larva begins to swim at lower water levels. Eventually it changes shape, acquires a pair of shells, six pairs of swimming legs, and a pair of antennae tipped with suckers. This "cypris" larva looks much like the adults of another group of crustaceans, the ostracods. Finally, guided by instinct to yield to gravity and to avoid light, it descends to the bottom ready to become an adult.

No one knows how many of the baby barnacles riding shoreward on the waves make a safe landing, how many fail in the quest for a clean, hard substratum. The settling down of a barnacle larva is not a haphazard process, but is performed only after a period of seeming deliberation. Biologists who have observed the act in the laboratory say the larvae "walk" about on the substratum for as long as an hour, pulling themselves along by the adhesive tips of the antennae, testing and rejecting many possible sites before they make a final choice. In nature they probably drift along in the currents for many days, coming down, examining the bottom at hand, then drifting on to another.

What are the conditions this infant creature requires? Probably it finds rock surfaces that are rough and pitted better than very smooth ones; probably it is repelled by a slimy film of microscopic plants, or even sometimes by the presence of hydroids or large algae. There is some reason to believe it may be drawn to existing colonies of barnacles perhaps through mysterious chemical attraction, detecting substances released by the adults and following these paths to the colony. Somehow, suddenly and irrevocably, the choice is made and the young barnacle cements itself to the chosen surface. Its tissues undergo a complete and drastic reorganization comparable to the metamorphosis of the larval butterfly. Then from an almost shapeless mass, the rudiments of the shell appear, the head and appendages are molded, and within twelve hours the complete cone of the shell, with all its plates delineated, has been formed.

Within its cup of lime the barnacle faces a dual growth problem. As a crustacean enclosed in a chitinous shell, the animal itself must periodically shed its unyielding skin so that its body may enlarge. Difficult as it seems, this feat is successfully accomplished, as I am reminded many times each summer. Almost every container of sea water that I bring up from the shore is flecked with white semitransparent objects, gossamer-fine, like the discarded garments of some very small fairy creature. Seen under the microscope, every detail of structure is perfectly represented. Evidently the barnacle accomplishes its withdrawal from the old skin with incredible neatness and thoroughness. In the little cellophane-like replicas I can count the joints of the appendages; even the bristles, growing at the bases of the joints, seem to have been slipped intact out of their casings.

The second problem is that of enlarging the hard cone to accommodate the growing body. Just how this is done no one seems to be sure, but probably there is some chemical secretion to dissolve the inner layers of the shell as new material is added on the outside.

Unless its life is prematurely ended by an enemy, a rock barnacle is likely to live about three years in the middle and lower tidal zones, or five years near the upper tidal levels. It can withstand high temperatures as rocks absorb the heat of the summer sun. Winter cold in itself is not harmful, but grinding ice may scrape the rocks clean. The pounding of the surf is part of the normal life of a barnacle; the sea is not its enemy.

When, through the attacks of fish, predatory worms, or snails, or through natural causes, the barnacle's life comes to an end, the shells remain attached to the rocks. These become shelter for many of the minute beings of the shore. Besides the baby periwinkles that regularly live there, the little tide-pool insects often hurry into these shelters if caught by the rising tide. And lower on the shore, or in tide pools, the empty shells are likely to house young anemones, tube worms, or even new generations of barnacles.

The chief enemy of the barnacle on these shores is a brightly colored carnivorous marine snail, the dog whelk. Although it preys also on mussels and even occasionally on periwinkles, it seems to prefer barnacles to all other food, probably because they are more easily eaten. Like all snails, the whelk possesses a radula. This is not used, periwinkle fashion, to scrape the rocks, but to drill a hole in any hard-shelled prey. It can then be pushed through the hole it has made, to reach and consume the soft parts within. To devour a barnacle, however, the whelk need only envelop the cone within its fleshy foot and force the valves open. It also produces a secretion that may have a narcotic effect. This is a substance called purpurin. In ancient times the secretion of a related snail in the Mediterranean was the source of the dye Tyrian purple. The pigment is an organic compound of bromine that changes in air to form a purple coloring matter.

Although violent surf excludes them, the dog whelks appear in numbers on most open shores, working up high into the zone of the barnacles and mussels. By their voracious feeding they may actually alter the balance of life on the shore. There is a story, for example, about an area where the whelks had reduced the number of barnacles so drastically that mussels came in to fill the vacant niche. When the whelks could find no more barnacles they moved over to the mussels. At first they were clumsy, not knowing how to eat the new food. Some spent futile days boring holes in empty shells; others climbed into empty shells and bored from inside. In time, however, they adjusted to the new prey and ate so many mussels that the colony began to dwindle. Then the barnacles settled anew on the rocks and in the end the snails returned to them.

In the middle stretches of shore and even down toward the low-tide line the whelks live under the dripping curtains of weed hanging down from the rock walls, or within the turf of Irish moss or among the flat, slippery fronds of the red seaweed, dulse. They cling to the under sides of overhanging ledges or gather in deep crevices where salt water drips from weeds and mussels, and little streams of water trickle over the floor. In all such places the whelks collect in numbers to pair and lay their eggs in straw-colored containers, each about the size and shape of a grain of wheat and tough as parchment. Each capsule stands alone, attached by its own base to the substratum, but usually they are crowded so closely together that they form a pattern or mosaic.

A snail takes about an hour to make one capsule but seldom completes more than 10 in twenty-four hours. It may produce as many as 245 in a season. Although a single capsule may contain as many as a thousand eggs, most of these are unfertilized nurse eggs that serve as food for the developing embryos. On maturing, the capsules become purple, being colored by the same chemical purpurin that is secreted by the adult. In about four months embryonic life is completed, and from 15 to 20 young dog whelks emerge from the capsule. Newly hatched young seldom if ever are found in the zone where the adults live, although the capsules are deposited and development takes place there. Apparently the waves carry the young snails down to low-tide level or below. Probably many are washed out to sea and lost, but the survivors are to be found at low water. They are very small—about one-sixteenth of an inch high—and are feeding on the tube worm, Spirorbis. Apparently the tubes of these worms are easier to penetrate than the cones of even very small barnacles. Not until the dog whelk is about one-fourth or three-eighths of an inch high does it migrate higher on the shore and begin feeding on barnacles.

Down in the middle sections of the shore the limpets become abundant. They appear sprinkled over the open rock surfaces, but most live numerously in shallow tide pools. A limpet wears a simple cone of shell the size of a fingernail, unobtrusively mottled with soft browns and grays and blues. It is one of the most ancient and primitive types of snails, and yet the primitiveness and the simplicity are deceptive. The limpet is adapted with beautiful precision to the difficult world of the shore. One expects a snail to have a coiled shell; the limpet has instead a flattened cone. The periwinkles, which have coiled shells, are often rolled around by the surf unless they have hidden themselves securely in crevices or under weeds. The limpet merely presses its cone against the rocks and the water slides over the sloping contours without being able to get a grip; the heavier the waves, the more tightly they press it to the rocks. Most snails have an operculum to shut out enemies and keep moisture inside; the limpet has one in infancy and then discards it. The fit of the shell to the substratum is so close that an operculum is unnecessary; moisture is retained in a little groove that runs around just inside the shell, and the gills are bathed in a small sea of their own until the tide returns.

Ever since Aristotle reported that limpets leave their places on the rocks and go out to feed, people have been recording facts about their natural history. Their supposed possession of a sort of homing sense has been widely discussed. Each limpet is said to possess a "home" or spot to which it always returns. On some types of rock there may be a recognizable scar, either a discoloration or a depression, to which the contours of the shell have become precisely fitted. From this home the limpet wanders out on the high tides to feed, scraping the small algae off the rocks by licking motions of the radula. After an hour or two of feeding it returns by approximately the same path, and settles down to wait out the period of low water.

Many nineteenth-century naturalists tried unsuccessfully to discover by experiment the nature of the sense involved and the organ in which the homing sense resides, much as modern scientists try to find a physical basis for the homing abilities of birds. Most of these studies dealt with the common British limpet, Patella, and although no one could explain how the homing instinct worked, there seemed to be little doubt in anyone's mind that it did work, and with great precision.

In recent years American scientists have investigated the matter with statistical methods, and some have concluded that Pacific coast limpets do not "home" very well at all. (No careful studies of homing have been made among New England limpets.) Other recent work in California, however, supports the homing theory. Dr. W. G. Hewatt marked a large number of limpets and their homes with identifying numbers. He found that on each high tide all the limpets left home, wandered about for some two and a half hours, then returned. The direction of their excursions changed from tide to tide, but they always returned to the home spot. Dr. Hewatt tried filing a deep groove across one animal's homeward path. The limpet halted on the edge of the groove and spent some time confronting this dilemma, but on the next tide it moved around the edge of the groove and returned home. Another limpet was taken about nine inches from its home and the edges of its shell were filed smooth. It was then released on the same spot. It returned to its home, but presumably the exact fit of shell to rock home had been destroyed by the filing and the next day the limpet moved about twenty-one inches away and did not return. On the fourth day it had taken up a new home and after eleven days it disappeared.

The limpet's relations with other inhabitants of the shore are simple. It lives entirely on the minute algae that coat the rocks with a slippery film, or on the cortical cells of larger algae. For either purpose, the radula is effective. The limpet scrapes the rocks so assiduously that fine particles of stone are found in its stomach; as the teeth of the radula wear away under hard use they are replaced by others, pushed up from behind. To the algal spores swarming in the water, ready to settle down and become sporelings and then adult plants, the limpets stand in the relation of enemy, since they keep the rocks scraped fairly clean where they are numerous. By this very act, however, they render a service to barnacles, making easier the attachment of their larvae. Indeed, the paths radiating out from a limpet's home are sometimes marked by a sprinkling of the starlike shells of young barnacles.

In its reproductive habits this deceptively simple little snail seems again to have defied exact observation. It seems certain, however, that the female limpet does not make protective capsules for her eggs in the fashion typical of many snails, but commits them directly to the sea. This is a primitive habit, followed by many of the simpler sea creatures. Whether the eggs are fertilized within the mother's body or while floating at sea is uncertain. The young larvae drift or swim for a time in the surface waters; the survivors then settle down on rocky surfaces and metamorphose from the larval to the adult form. Probably all young limpets are males, later transforming to females— a circumstance not at all uncommon among mollusks.

Like the animal life of this coast, the seaweeds tell a silent story of heavy surf. Back from the headlands and in bays and coves the rockweeds may grow seven feet tall; here on this open coast a seven-inch plant is a large one. In their sparse and stunted growth, the seaweed invaders of the upper rocks reveal the stringent conditions of life where waves beat heavily. In the middle and lower zones some hardy weeds have been able to establish themselves in greater abundance and profusion. These differ so greatly from the algae of quieter shores that they are almost a symbol of the wave-swept coast. Here and there the rocks sloping to the sea glisten with sheets composed of many individual plants of a curious seaweed, the purple laver. Its generic name, Porphyra, means "a purple dye." It belongs to the red algae, and although it has color variations, on the Maine coast it is most often a purplish brown. It resembles nothing so much as little pieces of brown transparent plastic cut out of someone's raincoat. In the thinness of its fronds it is like the sea lettuce, but there is a double layer of tissue, suggesting a child's rubber balloon that has collapsed so that the opposite walls are in contact. At the stem of the "balloon" Porphyra is attached strongly to the rocks by a cord of interwoven strands—hence its specific name, "umbilicalis." Occasionally it is attached to barnacles and very rarely it grows on other algae instead of directly on hard surfaces. When exposed at ebb tide under a hot sun, the laver may dry to brittle, papery layers, but the return of the sea restores the elastic nature of the plant, which, despite its seeming delicacy, allows it to yield unharmed to the push and pull of waves.

Down at the lower tidal levels is another curious weed— Leathesia, the sea potato. It grows in roughly globular form, its surface seamed and drawn into lobes, forming fleshy, amber-colored tubers that are any size up to an inch or two in diameter. Usually it grows around the fronds of moss or another seaweed, seldom if ever attaching directly to the rocks.

The lower rocks and the walls of low tide pools are thickly matted with algae. Here the red weeds largely supplant the browns that grow higher up. Along with Irish moss, dulse lines the walls of the pools, its thin, dull red fronds deeply indented so that they bear a crude resemblance to the shape of a hand. Minute leaflets sometimes haphazardly attached along the margins create a strangely tattered appearance. With the water withdrawn, the dulse mats down against the rocks, paper-thin layers piled one upon another. Many small starfish, urchins, and mollusks live within the dulse, as well as in the deeper growth of Irish moss.

Dulse is one of the algae that have a long history of usefulness to man, as food for himself and his domestic animals. According to an old book on seaweeds, it used to be said in Scotland that "he who eats of the Dulse of Guerdie and drinks of the wells of Kildingie will escape all maladies except black death." In Great Britain, cattle are fond of it and sheep wander down into the tidal zone at low water in search of it. In Scotland, Ireland, and Iceland people eat dulse in various ways, or dry it and chew it like tobacco; even in the United States, where such foods are usually ignored, it is possible to buy dulse fresh or dried in some coastal cities.

In the very lowest pools the Laminarias begin to appear, called variously the oarweeds, devil's aprons, sea tangles, and kelps. The Laminarias belong to the brown algae, which flourish in the dimness of deep waters and polar seas. The horsetail kelp lives below the tidal zone with others of the group, but in deep pools also comes over the threshold, just above the line of the lowest tides. Its broad, flat, leathery frond is frayed into long ribbons, its surface is smooth and satiny, and its color richly, glowingly brown.

The water in these deep basins is icy cold, filled with dusky, swaying plants. To look into such a pool is to behold a dark forest, its foliage like the leaves of palm trees, the heavy stalks of the kelps also curiously like the trunks of palms. If one slides his fingers down along such a stalk and grips just above the holdfast, he can pull up the plant and find a whole microcosm held within its grasp.

One of these laminarian holdfasts is something like the roots of a forest tree, branching out, dividing, subdividing, in its very complexity a measure of the great seas that roar over this plant. Here, finding secure attachment, are plankton-strainers like mussels and sea squirts. Small starfish and urchins crowd in under the arching columns of plant tissue. Predacious worms that have foraged hungrily during the night return with the daylight and coil themselves into tangled knots in deep recesses and dark, wet caverns. Mats of sponge spread over the holdfasts, silently, endlessly at their work of straining the waters of the pool. One day a larval bryozoan settles here, builds its tiny shell, then builds another and another, until a film of frosty lace flows around one of the rootlets of the seaweed. And above all this busy community, and probably unaffected by it, the brown ribbons of the kelp roll out into the water, the plant living its own life, growing, replacing torn tissues as best it may, and in season sending clouds of reproductive cells streaming into the water. As for the fauna of the holdfasts, the survival of the kelp is their survival. While it stands firm their little world holds intact; if it is torn away in a surge of stormy seas, all will be scattered and many will perish with it.

Among the animals almost always inhabiting the holdfasts of the tide-pool kelps are the brittle stars. These fragile echinoderms are well named, for even gentle handling is likely to cause them to snap off one or more arms. This reaction may be useful to an animal living in a turbulent world, for if an arm is pinned down under a shifting rock, the owner can break it off and grow a new one. Brittle stars move about rapidly, using their flexible arms not only in locomotion, but also to capture small worms and other minute sea life and carry them to their mouths.

The scale worms also belong to the holdfast community. Their bodies are protected by a double row of plates forming armament over the back. Under these large plates is an ordinary segmented worm, bearing laterally projecting tufts of golden bristles on each segment. There is a suggestion of primitiveness in the armor plate that is reminiscent of the quite unrelated chitons. Some of the scale worms have developed interesting relations with their neighbors. One of the British species always lives with burrowing animals, although it may change associates from time to time. When young, it lives with a burrowing brittle star, probably stealing its food. When older and larger, it moves into the burrow of a sea cucumber or the tube of the much larger, plumed worm, Amphitrite.

Often the holdfast grips one of the large horse mussels, which have heavy shells and may be four or five inches long. The horse mussel lives only in the deep pools or farther offshore; it is never found in the upper zones with the smaller blue mussel, and it occurs only on or among rocks, where its attachment is relatively secure. Sometimes it constructs a small nest or den as a refuge, using tough byssal threads spun in typical mussel fashion, with pebbles and shell fragments matted among the strands.

A small clam common in laminarian holdfasts is the rock-borer, which some English writers call the "red-nose" because of its red siphons. Ordinarily it is a boring form, living in cavities it excavates in limestone, clay, or concrete. Most of the New England rocks are too hard for boring, and so on this coast the clam lives in crusts of coralline algae or among the holdfasts of the kelp. On British coasts it is said to bore rocks that resist mechanical drills. And it does so without recourse to the chemical secretions some borers use, working entirely by repeated and endless mechanical abrasion with its sturdy shell.

The smooth and slippery fronds of the kelps support other populations, less abundant and less varied than those of the holdfasts. On the flat blades of the oarweeds, as well as on rock faces and under ledges, the golden-star tunicate, Botryllus, lays its spangled mats. Over a field of dark green gelatinous substance are sprinkled the little golden stars that mark the position of clusters of individual tunicates. Each starry cluster may consist of three to a dozen individual animals radiating around a central point; many clusters go to make up this continuous, encrusting mat, which may be six to eight inches long.

Beneath the surface beauty there is marvelous complexity of structure and function. Over each star there are infinitesimal disturbances in the water—little currents funneling down, one to each point of the star, and there being drawn in through a small opening.

One heavier, outward-moving current emerges from the center of the cluster. The indrawn currents bring in food organisms and oxygen, and the outflowing current carries away the metabolic waste products of the group.

At first glance a colony of Botryllus may seem no more complex than a mat of encrusting sponge. In actual fact, however, each of the individuals comprising the colony is a highly organized creature, in structure almost identical with such solitary ascidians as the sea grape and the sea vase, found so abundantly on wharves and sea walls. The individual Botryllus, however, is only one-sixteenth to one-eighth of an inch long.

One of these entire colonies, comprising perhaps hundreds of star clusters (and so perhaps a thousand or more individuals), may arise from a single fertilized ovum. In the parent colony, eggs are formed early in the summer, are fertilized, and begin their development while remaining within the parental tissues. (Each individual Botryllus produces both eggs and sperms, but since in any one animal they mature at different times, cross fertilization is insured, the spermatozoa being carried in the sea water and drawn in along with the water currents.) Presently the parent releases minute larvae shaped like tadpoles, with long, swimming tails. For perhaps an hour or two such a larva drifts and swims, then settles down on some ledge or weed and makes itself fast. Soon the tissues of the tail are absorbed and all suggestion of ability to swim is lost. Within two days the heart begins to beat in that curious tunicate rhythm—first driving the blood in one direction, pausing briefly, then reversing the direction of the flow. After nearly a fortnight, this small individual has completed the formation of its own body and begins to bud off other individuals. These, in turn, bud off others. Each new creature has its separate opening for the intake of water, but all retain connections with a central vent for the outflow of wastes. When the individuals clustered around this common opening become too crowded, one or more newly formed buds are pushed out into the surrounding mat of gelatinous tissue, where they begin new star clusters. In this way the colony spreads.

The intertidal zone is sometimes invaded by a deep-water laminarian, the sea colander. It is a representative of those brown seaweeds that flourish in the cold waters of the Arctic, and has come down from Greenland as far as Cape Cod. Its appearance is strikingly different from that of the sea moss and horsetail kelp among which it sometimes appears. The wide frond is pierced by innumerable perforations; these are foreshadowed in the young plant by conical papillae, which later break through to form the perforations.

Beyond the rims of the lowest pools, growing on the rock walls that slope away steeply into deep water, is another laminarian seaweed, Alaria, the winged kelp, called the murlin in Great Britain. Its long, ruffled, streaming fronds rise with each surge and fall as the water pours away seaward. The fertile pinnae, in which the reproductive cells mature, are borne at the base of the frond, for in a plant so exposed to violent surf this location is safer than the tips of the main blade. (In the rockweeds, living higher on the shore and less subject to savage wave action, the reproductive cells are formed at the tips of the fronds.) Almost more than any of the other seaweeds, Alaria is a plant conditioned to constant pounding by the waves. Standing on the outermost point that gives safe footing, one can see its dark ribbons streaming out into the water, being tugged and tossed and pounded. The larger and older plants become much frayed and worn, the margins of the blade splitting or the tip of the midrib being worn off. By such concessions the plant saves some of the strain on its holdfasts. The stipe can withstand a relatively enormous pull, but severe storms tear away many plants.

Still farther down, one can sometimes and in some places get a glimpse of the dark, mysterious forests of the kelps, where they go down into deep water. Sometimes these giant kelps are cast ashore after a storm. They have a stiff, strong stipe from which the long ribbon of the frond extends. The sea tangle or sugar kelp, Laminaria saccharina, has a stipe up to 4 feet long, supporting a relatively narrow frond (6 to 18 inches wide) that may extend out and upward into the sea as much as 30 feet. The margin is greatly frilled and a powdery white substance (mannitol, a sugar) forms on the dried fronds. The long-stalked laminaria (Laminaria longicruris) has a stem comparable to the trunk of a small tree, being 6 to 12 feet long. The frond is up to 3 feet wide and 20 feet long, but may sometimes be shorter than the stipe.

The stands of sea tangles and long-stalked laminarias are, in their way, an Atlantic counterpart of the great submarine jungles of the Pacific, where the kelps rise like giant forest trees, 150 feet from the floor of the sea to the surface.

On all rocky coasts, this laminarian zone just below low water has been one of the least-known regions of the sea. We know little about what lives here throughout the year. We do not know whether some of the forms that disappear from the intertidal area in winter may merely move down into this zone. And perhaps some of the species we think have died out in a particular region, perhaps because of temperature changes, have gone down among the Laminarias. The area is obviously difficult to explore, with heavy seas breaking there most of the time. Such an area on the west coast of Scotland was, however, explored by helmet divers working with the British biologist, J. A. Kitching. Below the zone occupied by Alaria and the horsetail kelp, from about two fathoms below low water and beyond, the divers moved through a dense forest of the larger Laminarias. From the vertical stipes an immense canopy of fronds was spread above their heads. Although the sun shone brightly at the surface, the divers were almost in darkness as they pushed through this forest. Between three and six fathoms below low water of the spring tides the forest opened out, so that the men could walk between the plants without great difficulty. There the light was stronger, and through misty water they could see this more open "park" extending farther down the sloping floor of the sea. Among the holdfasts and stipes of the laminarias, as among the roots and trunks of a terrestrial forest, was a dense undergrowth, here formed of various red algae. And as small rodents and other creatures have their dens and runways under the forest trees, so a varied and abundant fauna lived on and among the holdfasts of the great seaweeds.

In quieter waters, protected from the heavy surf of coasts that face the open ocean, the seaweeds dominate the shore, occupying every inch of space that the conditions of tidal rise and fall allow them and by the sheer force of abundant and luxuriant growth forcing other shore inhabitants to accommodate to their pattern.

Although the same bands of life are spread between the tide lines whether the coast be open or sheltered, in their relative development the zones vary greatly on the two types of shore.

Above the high-tide line there is little change and on the shores of bays and estuaries, as elsewhere, the microplants blacken the rocks and the lichens come down and tentatively approach the sea. Below high water of spring tides, pioneering barnacles trace occasional white streaks in token occupation of the zone they dominate on open coasts. A few periwinkles graze on the upper rocks. But on sheltered coasts the whole band of shore marked out by the tides of the moon's quarters is occupied by a swaying submarine forest, sensitive to the movements of the waves and the tidal currents. The trees of the forest are the large sea weeds known as the rockweeds or sea wracks, stout of form and rubbery of texture. Here all other life exists within their shelter—a shelter so hospitable to small things needing protection from drying air, from rain, and from the surge of the running tides and the waves, that the life of these shores is incredibly abundant.

When covered at high tide, the rockweeds stand erect, rising and swaying with a life borrowed from the sea. Then, to one standing at the edge of a flooding tide, the only sign of their presence may be a scattering of dark patches on the water close inshore, where the tips of the weeds reach up to the surface. Down below those floating tips small fishes swim, passing between the weeds as birds fly through a forest, sea snails creep along the fronds, and crabs climb from branch to branch of the swaying plants. It is a fantastic jungle, mad in a Lewis Carroll sort of way. For what proper jungle, twice every twenty-four hours, begins to sag lower and lower and finally lies prostrate for several hours, only to rise again? Yet this is precisely what the rockweed jungles do. When the tide has retreated from the sloping rocks, when it has left the miniature seas of the tide pools, the rockweeds lie flat on the horizontal surfaces in layer above layer of sodden, rubbery fronds. From the sheer rock faces they hang down in a heavy curtain, holding the wetness of the sea, and nothing under their protective cover ever dries out.

By day the sunlight filters through the jungle of rockweeds to reach its floor only in shifting patches of shadow-flecked gold; by night the moonlight spreads a silver ceiling above the forest—a ceiling streaked and broken by the flowing tide streams; beneath it the dark fronds of the weeds sway in a world unquiet with moving shadows.

But the flow of time through this submarine forest is marked less by the alternation of light and darkness than by the rhythm of the tides. The lives of its creatures are ruled by the presence or absence of water; it is not the fall of dusk or the coming of dawn but the turn of the tide that brings transforming change to their world.

As the tide falls the tips of the weeds, lacking support, float out horizontally across the surface. Then the cloud shadows darken and a deepening gloom settles over the floor of the forest. As the overlying layer of water thins and gradually drains away, the weeds, still stirring, still responsive to each pulsation of the tide, drift closer to the rock floor and finally lie prostrate upon it, all their life and movement in abeyance.

By day an interval of quiet settles over the jungles of the land, when the hunters lie in their dens, and the weak and the slow hide from the daylight; so on the shore a waiting lull comes with every ebbing of the tide.

The barnacles furl their nets and swing shut the twin doors that exclude the drying air and hold within the moisture of the sea. The mussels and the clams withdraw their feeding tubes or siphons and close their shells. Here and there a starfish, having invaded the forest from below on the previous high tide and incautiously lingered, still clasps a mussel within its sinuous arms, gripping the shells with the sucker-tipped ends of scores of slender tube feet. Pushing under and among the horizontal fronds of the weed, as a man would make his way with difficulty through trees blown down by a storm, a few crabs are active, digging their little slanting pits to expose the clams buried in the mud. Then they crack away pieces of shell with their heavy claws, while they hold the clam in the tips of the walking legs.

A few hunters and scavengers come down from the upper tidelands. The little gray-cloaked tide-pool insect, Anurida, wanders down from the upper shore and scurries over the rock floor, hunting out mussels with gaping shells, or dead fish, or fragments of crabs left by gulls. Crows walk about over the weeds; they sort them over strand by strand until they find a periwinkle hidden in the weed, or clinging to a rock that lies under the sodden cloak of the algae. Then the crow holds the shell in the strong toes of one foot, while with its beak it deftly extracts the snail.

The pulse of the returning tide at first beats gently. The advance during the beginning of the six-hour rise to high-water mark is slow, so that in two hours only a quarter of the intertidal zone has been covered. Then the pace of the water quickens. For the next two hours the tidal currents are stronger and the rising waters advance twice as far as in the first period; then again the tide slackens its pace for a leisurely advance over the upper shore. The rockweeds, covering the middle band of shore, receive the shock of heavier waves than the relatively bare shore above, yet their cushioning effect is so great that the animals that cling to them or live on the rock floor below them are far less affected by the surf than those of the upper rocks, or those of the zone below which experience all the heavy drag from the backwash of waves that break as the tide is advancing rapidly over the middle shore.

Darkness brings the jungles of the land to life, but the night of the rockweed jungles is the time of the rising tide, when water pours in under the masses of weed, stirring out of their low-tide quiescence all the inhabitants of this forest.

As the water from the open sea floods the floor of the weed jungles, shadows flicker again above the ivory cones of the barnacles as the almost invisible nets reach out to gather what the tide has brought. The shells of clams and mussels again open slightly and little vortices of water are drawn down, funneling into the complex straining mechanisms within the shellfish all the little spheres of marine vegetables that are their food.

Nereid worms emerge from the mud and swim off to other hunting grounds; if they are to reach them they must elude the fishes that have come in with the tide, for on the flood tide the rockweed forests become one with the sea and with its hungry predators.

Shrimp flicker in and out through the open spaces of the forest; they seek small crustaceans, baby fish, or minute bristle worms, but in their turn are pursued by following fish. Starfish move up from the great meadows of sea moss lower on the shore, hunting the mussels that grow on the floor of the forest.

The crows and the gulls are driven out of the tidelands. The little gray, velvet-cloaked insects move up the shore or, finding a secure crevice, wrap themselves in a glistening blanket of air to wait for the falling of the tide.

The rockweeds that create this intertidal forest are descendants of some of the earth's most ancient plants. Along with the great kelps lower on the shore, they belong to the group of brown seaweeds, in which the chlorophyll is masked by other pigments. The Greek name for the brown algae—the Phaeophyceae— means "the dusky or shadowy plants." According to some theories, they arose in that early period when the earth was still enveloped in heavy clouds and illuminated only by feeble rays of sunlight. Even today the brown seaweeds are plants of dim and shadowed places—the deep submarine slopes where giant kelps form dusky jungles and the dark rock ledges from which the oarweeds send their long ribbons streaming into the tides. And the rockweeds that grow between the tide lines do so on northern coasts, visited often by cloud and fog. Their rare invasions of the sunny tropics are accomplished under a protective cover of deep water.

The brown algae may have been the first of the sea plants to colonize the shore. They learned to adjust themselves to alternating periods of submersion and exposure on ancient coastlines swept by strong tides; they came as close to a land existence as they could without actually leaving the tidal zone.

One of the modern rockweeds, the channeled wrack of European shores, lives at the extreme upper edge of the tidelands. In some places its only contact with the sea is an occasional drenching with spray. In sun and air its fronds become blackened and crisp so that one would think it had surely been killed, but with the return of the sea its normal color and texture are restored.

The channeled wrack does not grow on the American Atlantic coast, but there a related plant, the spiral wrack, comes almost as far out of the sea. It is a weed of low growth, whose short sturdy fronds end in turgid, rough-textured swellings. Its heaviest growth is above the high-water mark of the neap tides, so of all the rockweeds it lives closest inshore or nearest the water line of exposed ledges. Although it spends nearly three-fourths of its life out of water, it is a true seaweed and its splashes of orange-brown color on the upper shore are a symbol of the threshold of the sea.

These plants, however, are but the outlying fringe of the intertidal forest, which is an almost pure stand of two other rockweeds—the knotted wrack and the bladder wrack. Both are sensitive indicators of the force of the surf. The knotted wrack can live in profusion only on shores protected from heavy waves, and in such places is the dominant weed. Back from the headlands, on the shores of bays and tidal rivers where surf and tidal surge are subdued by remoteness from the open sea, the knotted wrack may grow taller than the tallest man, though its fronds are slender as straws. The long surge of the swell in sheltered water places no great strain on its elastic strands. Swellings or vesicles on the main stems or fronds contain oxygen and other gases secreted by the plant; these act as buoys when the weeds are covered by the tide. The bladder wrack has greater tensile strength and so can endure the sharp tugging and pulling of moderately heavy surf. Although it is much shorter than the knotted wrack it also needs the help of air bladders to rise in the water. In this species the bladders are paired, one of each pair on either side of the strong midrib; the bladders, however, may fail to develop where the plants are subjected to much pounding by surf, or when they grow at the lower levels of the tidal zone. At some seasons the ends of the branches of this wrack swell into bulbous, almost heart-shaped structures; from these the reproductive cells are liberated.

The sea wracks have no roots, but instead grip the rocks by means of a flattened, disc-like expansion of their tissues. It is almost as though the base of each weed melted a little, spreading over the rock and then congealing, thereby creating a union so firm that only the thundering seas of a very heavy storm, or the grinding of shore ice, can tear away the plants. The seaweeds do not have a land plant's need of roots to extract minerals from the soil, for they are bathed almost continuously by the sea and so live within a solution of all the minerals they need for life. Nor do they need the rigid supporting stem or trunk by which a land plant reaches upward into sunlight—they have only to yield themselves to the water. And so their structure is simple—merely a branching frond arising from the holdfast, with no division into roots and stems and leaves.

Looking at the prostrate, low-tide forests of the rockweeds that cover the shore with a many-layered blanket, one would suppose that the plants must spring from every available inch of rock surface. But actually the forest, when it rises and comes to life with the flooding tide, is fairly open and sprinkled with clearings. On my own shore in Maine, where the tides rise and fall over a wide expanse of intertidal rock, and the knotted wrack spreads its dark blanket between the high and low waters of the neap tides, the areas of open rock around the holdfast of each plant are sometimes as much as a foot in diameter. From the middle of such a clearing the plant rises, its fronds dividing repeatedly, until the upper branchings extend out over an area several feet across.

Far below, at the base of the fronds that swing with the undulation of the passing waves, the rocks are stained with vivid hues, painted in crimson and emerald by the activities of sea plants so minute that even in their thousands they seem but part of the rock, a surface revelation of jewel tones within. The green patches are growths of one of the green algae. The individual plants are so small that only a strong lens could reveal their identity—lost, as individual blades of grass are lost in the lush expanse of a meadow, in the spreading verdant stain created by the mass. Amid the green are other patches of a rich and intensely glowing red, and again the growth is not separable from the mineral floor. It is a creation of one of the red seaweeds, a form that secretes lime in thin and closely adhering crusts over the rocks.

Against this background of glowing color the barnacles stand out with sharp distinctness, and in the clear water that pours through the forest like liquid glass, their cirri flicker in and out-extending, grasping, withdrawing, taking from the inpouring tides those minute atoms of life that our eyes cannot see. Around the bases of small wave-rounded boulders the mussels lie as though at anchor, held by gleaming lines spun by their own tissues. Their paired blue shells stand a little apart, the space between them revealing pale brown tissues with fluted edges.

Some parts of the forest are less open. In these the clumps of rockweeds rise from a short turf or undergrowth consisting chiefly of the flat fronds of Irish moss, with sometimes dark mats of another plant with the texture of Turkish toweling. And like a tropical jungle with its orchids, this sea forest has the counterpart of airplants in the epiphytic tufts of a red seaweed that grows on the fronds of the knotted wrack. Polysiphonia seems to have lost—or perhaps it never had—the ability to attach directly to the rocks and so its dark red balls of finely divided fronds cling to the wracks, and by them are lifted up into the water.

In the areas between the rocks and under loose boulders a substance that is neither sand nor mud has accumulated. It consists of minute and water-ground bits of the remains of sea creatures—the shells of mollusks, the spines of sea urchins, the opercula of snails. Clams live in pockets of this soft substance, digging down until they are buried to the tips of their siphons. Around the clams the mud is alive with ribbon worms, thin as threads, scarlet of color, each a small hunter searching out minute bristle worms and other prey. Here also are the nereids, given the Latin name for sea nymph because of their grace and iridescent beauty. The nereids are active predators that leave their burrows at night to search for small worms, crustaceans, and other prey. In the dark of the moon certain species gather at the surface in immense spawning swarms. Curious legends have become associated with them. In New England the so-called clam worm, Nereis virens, often takes shelter in empty clam shells. Fishermen, accustomed to finding it thus, believe it is the male clam.

Crabs of thumbnail size live in the weed and come down to hunt in these areas. They are the young of the green crab; the adults live below the tide lines on this shore except when they come into the shelter of the weeds to molt. The young crabs search the mud pockets, digging out pits and probing for clams that are about their own size.

Clams, crabs, and worms are part of a community of animals whose lives are closely interrelated. The crabs and the worms are the active predators, the beasts of prey. The clams, the mussels, and the barnacles are the plankton feeders, able to live sedentary lives because their food is brought to them by each tide. By an immutable law of nature, the plankton feeders as a group are more numerous than those that prey on them. Besides the clams and other large species, the rockweeds shelter thousands of small beings, all of them busy with filtering devices of varying design, straining out the plankton of each tide. There is, for example, a small, plumed worm called Spirorbis. Seeing it for the first time, one would certainly say that it is no worm, but a snail, for it is a tube-builder, having learned some feat of chemistry that allows it to secrete about itself a calcareous shell or tube. The tube is not much larger than the head of a pin and is wound in a flat, closely coiled spiral of chalky whiteness, its form strongly suggesting some of the land snails. The worm lives permanently within the tube, which is cemented to weed or rock, thrusting out its head from time to time to filter food animals through the fine filaments of its crown of tentacles. These exquisitely delicate and filmy tentacles serve not only as snares to entangle food but as gills for breathing. Among them is a structure like a long-stemmed goblet; when the worm draws back into its tube the goblet or operculum closes the opening like a neatly fitted trap door.

The fact that the tube worms have managed to live in the intertidal zone for perhaps millions of years is evidence of a sensitive adjustment of their way of life, on the one hand to conditions within the surrounding world of the rockweeds, on the other to vast tidal rhythms linked with the movements of earth, moon, and sun.

In the inmost coils of the tube are little chains of beads wrapped in cellophane—or so they appear. There are about twenty beads in a chain. The beads are developing eggs. When the embryos have developed into larvae, the cellophane membranes rupture and the young are sent forth into the sea. By keeping the embryonic stages within the parental tube Spirorbis protects its young from enemies and assures that the infant worms will be in the intertidal zone when they are ready to settle. Their period of active swimming is short—at most an hour or so, and well contained within a single rising or falling of the tide. They are stout little creatures with bright red eye spots; perhaps the larval eyes help in locating a place for attachment but in any event they degenerate soon after the larva settles.

In the laboratory, under my microscope, I have watched the

larvae swimming about busily, all their little bristles whirring, then sometimes descending to the glass floor of their dish to bump it with their heads. Why and how do the infant tube worms settle in the same sort of place their ancestors chose? Apparently they make many trials, reacting more favorably to smooth surfaces than to rough, and displaying a strong instinct of gregariousness that leads them to settle by preference where others of their kind are already established. These tendencies help to keep the tube worms within their comparatively restricted world. There is also a response, not to familiar surroundings, but to cosmic forces. Every fortnight, on the moon's quarter, a batch of eggs is fertilized and taken into the brood chamber to begin its development. And at the same time the larvae that have been made ready during the previous fortnight are expelled into the sea. By this timing—this precise synchronizing with the phases of the moon—the release of the young always occurs on a neap tide, when neither the rise nor the fall of the water is of great extent, and even for so small a creature the chances of remaining within the rockweed zone are good.

Sea snails of the periwinkle tribe inhabit the upper branches of the weeds at high tide or take shelter under them when the tide is out. The orange and yellow and olive-green colors of their smoothly rounded, flat-topped shells suggest the fruiting bodies of the rockweeds, and perhaps the resemblance is protective. The smooth periwinkle, unlike the rough, is still an animal of the sea; the salty dampness it requires is provided by the wet and dripping fronds of the seaweeds when the tide is out. It lives by scraping off the cortical cells of the algae, seldom if ever descending to the rocks to feed on the surface film as related species do. Even in its spawning habits the smooth periwinkle is a creature of the rockweeds. There is no shedding of eggs into the sea, no period of juvenile drifting in the currents. All the stages of its life are lived in the rockweeds—it knows no other home.

Curious about the early stages of this abundant snail, I have gone down into my own rockweed forests on the summer low tides to search for them. Sorting over the prostrate wrack, examining its long strands for some signs of what I sought, I have occasionally been rewarded by discovering transparent masses of a substance like tough jelly, tightly adhering to the fronds. They averaged perhaps a quarter-inch long and half as wide. Within each mass 1 could see the eggs, round as bubbles, dozens of them embedded in the confining matrix. One such egg mass that I carried to the microscope contained a developing embryo within the membranes of each egg. They were clearly molluscan, but so undifferentiated that I could not have said what mollusk lay nascent within. In the cold waters of its home, about a month would intervene from the egg to the hatching stage, but in the warmer temperatures of the laboratory the remaining days of development were reduced to hours. The following day each sphere contained a tiny baby periwinkle, its shell completely formed, apparently ready to emerge and take up its life on the rocks. How do they hold their places there, I wondered, as the weeds sway in the tides and occasional storms send waves pounding in over the shore? Later in the summer there was at least a partial answer. I noticed that many of the air vesicles of the wracks bore little perforations, as though they had been chewed or punctured by some animal. I slit some of these vesicles carefully so that I might look inside. There, secure in a green-walled chamber, were the babies of the smooth periwinkle—from two to half a dozen of them sharing the refuge of a single vesicle, secure alike against storms and enemies.

Down near the low water of the neap tides the hydroid Clava spreads its velvet patches on the fronds of the knotted wrack and the bladder wrack. Rising from its point of attachment like a plant from its root clump, each cluster of tubular animals looks like nothing so much as a spray of delicate flowers, shading from pink to rose and fringed with petal-like tentacles, all nodding in the water currents as woodland flowers nod in a gentle wind. But the swaying movements are purposive ones by which the hydroid reaches into the currents for food. In its way it is a voracious little jungle beast, all its tentacles studded with batteries of stinging cells that can be shot into its victims like poisoned arrows. When, in their ceaseless movements, the tentacles come into contact with a small crustacean or worm or the larva of some sea creature, a shower of darts is released; the prey animal becomes paralyzed and is seized and conveyed to the mouth by the tentacles.

Each of these colonies now established on the wracks came from a little swimming larva that once settled there, shed the hairy cilia by which it swam, attached itself, and began to elongate into a little plantlike being. A crown of tentacles formed at its free end. In time, from the base of the tubular creature, a seeming root, or stolon, began to creep over the rockweed, budding off new tubes, each complete with mouth and tentacles. So all the numerous individuals of the colony originated in a single fertilized ovum that yielded the wandering larva.

In season, the plantlike hydroid must reproduce, but by a strange circumstance it cannot itself yield the germ cells that would give rise to new larvae, for it can reproduce only non-sexually, by budding. So there is a curious alternation of generations, found again and again in many members of the large coelenterate group to which the hydroids belong, by which no individual produces offspring that resemble itself, but each is like the grandparental generation. Just below the tentacles of an individual Clava the buds of the new generation are produced— the alternate generation that intervenes between colonies of hydroids. They are pendent clusters shaped like berries. In some species the berries, or medusa buds, would drop from the parent and swim away— tiny, bell-shaped things like minute jellyfish. Clava, however, does not release its medusae but keeps them attached. Pink buds are male medusae; purple ones are female. When they are mature, each sheds its eggs or sperm into the sea. When fertilized, the eggs begin to divide and through their development yield the little protoplasmic threads of larvae, which swim off through unknown waters to found some distant colonies.

During many days of midsummer, the incoming tides bring

the round opalescent forms of the moon jellies. Most of these are in the weakened condition that accompanies the fulfillment of their life cycle; their tissues are easily torn by the slightest turbulence of water, and when the tide carries them in over the rockweeds and then withdraws, leaving them there like crumpled cellophane, they seldom survive the tidal interval.

Each year they come, sometimes only a few at a time, sometimes in immense numbers. Drifting shoreward, their silent approach is unheralded even by the cries of sea birds, who have no interest in the jellyfish as food, for their tissues are largely water.

During much of the summer they have been drifting offshore, white gleams in the water, sometimes assembling in hundreds along the line of meeting of two currents, where they trace winding lines in the sea along these otherwise invisible boundaries. But toward autumn, nearing the end of life, the moon jellies offer no resistance to the tidal currents, and almost every flood tide brings them in to the shore. At this season the adults are carrying the developing larvae, holding them in the flaps of tissue that hang from the under surface of the disc. The young are little pear-shaped creatures; when finally they are shaken loose from the parent (or freed by the stranding of the parent on the shore), they swim about in the shallow water, sometimes swarms of them together. Finally they seek bottom and each becomes attached by the end that was foremost when it swam. As a tiny plantlike growth, about an eighth of an inch high and bearing long tentacles, this strange child of the delicate moon jelly survives the winter storms. Then constrictions begin to encircle its body, so that it comes to resemble a pile of saucers. In the spring these "saucers" free themselves one after another and swim away, each a tiny jellyfish, fulfilling the alternation of the generations. North of Cape Cod these young grow to their full diameter of six to ten inches by July; they mature and produce eggs and sperm cells by late July or August; and in August and September they begin to yield the larvae that will become the attached generation. By the end of October all of the season's jellyfish have been destroyed by storms, but their offspring survive, attached to the rocks near the low-tide line or on nearby bottoms offshore.

If the moon jellies are symbols of the coastal waters, seldom straying more than a few miles offshore, it is otherwise with the great red jellyfish, Cyanea, which in its periodic invasions of bays and harbors links the shallow green waters with the bright distances of the open sea. On fishing banks a hundred or more miles offshore one may see its immense bulk drifting at the surface as it swims lazily, its tentacles sometimes trailing for fifty feet or more. These tentacles spell danger for almost all sea creatures in their path and even for human beings, so powerful is the sting. Yet young cod, haddock, and sometimes other fishes adopt the great jellyfish as a "nurse," traveling through the shelterless sea under the protection of this large creature and somehow unharmed by the nettle-like stings of the tentacles.

Like Aurelia, the red jellyfish is an animal only of the summer seas, for whom the autumnal storms bring the end of life. Its offspring are the winter plantlike generation, duplicating in almost every detail the life history of the moon jelly. On bottoms no more than two hundred feet deep (and usually much less), little half-inch wisps of living tissue represent the heritage of the immense red jellyfish. They can survive the cold and the storms that the larger summer generation cannot endure; when the warmth of spring begins to dissipate the icy cold of the winter sea they will bud off the tiny discs that, by some inexplicable magic of development, grow in a single season into the adult jellyfish.

As the tide falls below the rockweeds, the surf of the sea's edge washes over the cities of the mussels. Here, within these lower reaches of the intertidal zone, the blue-black shells form a living blanket over the rocks. The cover is so dense, so uniform in its texture and composition, that often one scarcely realizes that this is not rock, but living animals. In one place the shells, unimaginable in number, are no more than a quarter of an inch long; in another the mussels may be several times as large. But always they are packed so closely together, neighbor against neighbor, that it is hard to see how any one of them can open its shells enough to receive the currents of water that bring its food. Every inch, every hundredth of an inch of space, has been taken over by a living creature whose survival depends on gaining a foothold on this rocky shore.

The presence of each individual mussel in this crowded assemblage is evidence of the achievement of its unconscious, juvenile purpose, an expression of the will-to-live embodied in a minute transparent larva once set adrift in the sea to find its own solid bit of earth for attachment, or to die.

The setting adrift takes place on an astronomical scale. Along the American Atlantic coast the spawning season of the mussels is protracted, extending from April into September. What induces a wave of spawning at any particular time is unknown, but it seems clear that the spawning of a few mussels releases chemical substances into the water, and that these react on all mature individuals in the area and set them to pouring their eggs and milt into the sea. The female mussels discharge the eggs in a continuing, almost endless stream of short little rod-like masses—hundreds, thousands, millions of cells, each potentially an adult mussel. One large female may release up to twenty-five million at a single spawning. In quiet water the eggs drift gently to the bottom, but in the normal conditions of surf or swiftly moving currents they are at once possessed by the sea and carried away.

Simultaneously with the outflow of eggs, the water has become cloudy with the milt poured into the water by the male mussels, the number of individual sperm cells defying all attempts at calculation. Dozens of them cluster about a single egg, pressing against it, seeking entrance. But one male cell, and one only, is successful. With the entrance of this first sperm cell, an instantaneous physical change takes place in the outer membranes of the egg, and from this moment it cannot again be penetrated by a spermatozoan.

After the union of the male and female nuclei, the division of the fertilized cell proceeds rapidly. In less than the interval between a high and a low tide, the egg has been transformed into a little ball of cells, propelling itself through the water with glittering hairs, or cilia. In about twenty-four hours, it has assumed an odd, top-shaped form that is common to the larvae of all young mollusks and annelid worms. A few days more and it has become flattened and elongated and swims rapidly by vibrations of a membrane called the velum; it crawls over solid surfaces, and senses contact with foreign objects. Its journey through the sea is far from being a solitary one; in a square meter of surface over a bed of adult mussels there may be as many as 170,000 swimming larvae.

The thin larval shell takes form, but soon it is replaced by another, double-valved as in adult mussels. By this time the velum has disintegrated, and the mantle, foot, and other organs of the adult have begun their development.

From early summer these tiny shelled creatures live in prodigious numbers in the seaweeds of the shore. In almost every bit of weed I pick up for microscopic examination I find them creeping about, exploring their world with the long tubular organ called the foot, which bears an odd resemblance to the trunk of an elephant. The infant mussel uses it to test out objects in its path, to creep over level or steeply sloping rocks or through seaweeds, or even to walk on the under side of the surface film of quiet water. Soon, however, the foot assumes a new function: it aids in the work of spinning the tough silken threads that anchor a mussel to whatever offers a solid support and insurance against being washed away in the surf.

The very existence of the mussel fields of the low-tide zone is evidence that this chain of circumstances has proceeded unbroken to its consummation untold millions upon millions of times. Yet, for every mussel surviving upon the rocks, there must have been millions of larvae whose setting forth into the sea had a disastrous end. The system is in delicate balance; barring catastrophe, the forces that destroy neither outweigh nor are outweighed by those that create, and over the years of a man's life, as over the ages of recent geologic time, the total number of mussels on the shore probably has remained about the same.

In much of this low-water area the mussels live in intimate association with one of the red seaweeds, Gigartina, a plant of low-growing, bushy form and almost cartilaginous texture. Plants and mussels unite inseparably to form a tough mat. Very small mussels may grow about the plants so abundantly as to obscure their basal attachment to the rocks. Both the stems and the repeatedly subdivided branches of the seaweed are astir with life, but with life on so small a scale that the human eye can see its details only with the aid of a microscope.

Snails, some with brightly banded and deeply sculptured shells, crawl along the fronds, browsing on microscopic vegetable matter. Many of the basal stems of the weed are thickly encrusted with the bryozoan sea lace, Membranipora; from all its compartments the minute, be-tentacled heads of the resident creatures are thrust out. Another bryozoan of coarser growth, Flustrella, also forms mats investing the broken stems and stubble of the red weed, the substance of its own growth giving such a stem almost the thickness of a pencil. Rough hairs or bristles protrude from the mat, so that much foreign matter adheres to it. Like the sea laces, however, it is formed of hundreds of small, adjacent compartments. From one after another of these, as I watch through my microscope, a stout little creature cautiously emerges, then unfurls its crown of filmy tentacles as one would open an umbrella. Threadlike worms creep over the bryozoan, winding among the bristles like snakes through coarse stubble. A tiny, cyclopean crustacean, with one glittering ruby eye, runs ceaselessly and rather clumsily over the colony, apparently disturbing the inhabitants, for when one of them feels the touch of the blundering crustacean it quickly folds its tentacles and withdraws into its compartment.

In the upper branches of this jungle formed by the red weed, there are many nests or tubes occupied by amphipod crustaceans known as Amphithoe. These small creatures have the appearance of wearing cream-colored jerseys brightly splotched with brownish red; in each goatlike face are set two conspicuous eyes and two pairs of hornlike antennae. The nests are as firmly and skillfully constructed as a bird's but are subject to far more continuous use, for these amphipods are weak swimmers and ordinarily seem loath to leave their nests. They lie in their snug little sacs, often with the heads and upper parts of their bodies protruding. The water currents that pass through their seaweed home bring them small plant fragments and thus solve the problem of subsistence.

For most of the year Amphithoe lives singly, one to a nest. Early in the summer the males visit the females (who greatly outnumber them) and mating occurs within the nest. As the young develop the mother cares for them in a brood-pouch formed by the appendages of her abdomen. Often, while carrying her young, she emerges almost completely from her nest and vigorously fans currents of water through the pouch.

The eggs yield embryos, the embryos become larvae; but still the mother holds and cares for them until their small bodies have so developed that they are able to set forth into the seaweeds, to spin their own nests out of the fibers of plants and the silken threads mysteriously fashioned in their own bodies, and to feed and fend for themselves.

As her young become ready for independent life, the mother shows impatience to be rid of the swarm in her nest. Using claws and antennae, she pushes them to the rim and with shoves and nudges tries to expel them. The young cling with hooked and bristled claws to the walls and doorway of the familiar nursery. When finally thrust out they are likely to linger nearby; when the mother incautiously emerges, they leap to attach themselves to her body and so be drawn again into the security of their accustomed nest, until maternal impatience once more becomes strong.

Even the young just out of the brood-sac build their own nests and enlarge them as their growth requires. But the young seem to spend less time than the adults do inside their nests, and to creep about more freely over the weeds. It is common to see several tiny nests built close to the home of a large amphipod; perhaps the young like to stay close to the mother even after they have been ejected from her nest.

At low tide the water falls below the rockweeds and the mussels and enters a broad band clothed with the reddish-brown turf of the Irish moss. The time of its exposure to the atmosphere is so brief, the retreat of the sea so fleeting, that the moss retains a shining freshness, a wetness, and a sparkle that speak of its recent contact with the surf. Perhaps because we can visit this area only in that brief and magical hour of the tide's turning, perhaps because of the nearness of waves breaking on rocky rims, dissolving in foam and spray, and pouring seaward again to the accompaniment of many water sounds, we are reminded always that this low-tide area is of the sea and that we are trespassers.

Here, in this mossy turf, life exists in layers, one above another; life exists on other life, or within it, or under it, or above it. Because the moss is low-growing and branches profusely and intricately, it cushions the living things within it from the blows of the surf, and holds the wetness of the sea about them in these brief intervals of the low ebbing of the tide. After I have visited the shore and then at night have heard the surf trampling in over these moss-grown ledges with the heavy tread of the fall tides, I have wondered about the baby starfish, the urchins, the brittle stars, the tube-dwelling amphipods, the nudibranchs, and all the other small and delicate fauna of the moss; but I know that if there is security in their world it should be here, in this densest of intertidal jungles, over which the waves break harmlessly.

The moss forms so dense a covering that one cannot see what is beneath without intimate exploration. The abundance of life here, both in species and individuals, is on a scale that is hard to grasp. There is scarcely a stem of Irish moss that is not completely encased with one of the bryozoan sea mats—the white lacework of Membranipora or the glassy, brittle crust of Microporella. Such a crust consists of a mosaic of almost microscopic cells or compartments, arranged in regular rows and patterns, their surfaces finely sculptured. Each cell is the home of a minute, tentacled creature. By a conservative guess, several thousand such creatures live on a single stem of moss. On a square foot of rock surface there are probably several hundred such stems, providing living space for about a million of the bryozoans. On a stretch of Maine shore that the eye can take in at a glance, the population must run into the trillions for this single group of animals.

But there are further implications. If the population of the sea laces is so immense, that of the creatures they feed upon must be infinitely greater. A bryozoan colony acts as a highly efficient trap or filter to remove minute food animals from the sea water. One by one, the doors of the separate compartments open and from each a whorl of petal-like filaments is thrust out. In one moment the whole surface of the colony may be alive with crowns of tentacles swaying like flowers in a windswept field; the next instant, all may have snapped back into their protective cells and the colony is again a pavement of sculptured stone. But while the "flowers" sway over the stone field each spells death for many beings of the sea, as it draws in the minute spheres and ovals and crescents of the protozoans and the smallest algae, perhaps also some of the smallest of crustaceans and worms, or the larvae of mollusks and starfish, all of which are invisibly present in this mossy jungle, in numbers like the stars.

Larger animals are less numerous but still impressively abundant. Sea urchins, looking like large green cockleburs, often lie deep within the moss, their globular bodies anchored securely to the underlying rock by the adhesive discs of many tube feet. The ubiquitous common periwinkles, in some curious way unaffected by the conditions that confine most intertidal animals to certain zones, live above, within, and below the moss zone. Here their shells lie about over the surface of the weed at low tide; they hang heavily from its fronds, ready to drop at a touch.

And young starfish are here by the hundred, for these meadows of moss seem to be one of the chief nurseries for the starfish of northern shores. In the fall almost every other plant shelters quarter-inch and half-inch sizes. In these youthful starfish there are color patterns that become obliterated in maturity. The tube feet, the spines, and all the other curious epidermal outgrowths of these spiny-skinned creatures are large in proportion to the total size and have a clean perfection of form and structure.

On the rocky floor among the plant stems lie the infant stars. They are white insubstantial specks, in size and delicate beauty like snowflakes. There is an obvious newness about them, proclaiming that they have undergone their metamorphosis from the larval form to the adult shape only recently.

Perhaps it was on these very rocks that the swimming larvae, completing their period of life in the plankton, came to rest, attaching themselves firmly and becoming for a brief period sedentary animals. Then their bodies were like blown glass from which slender horns projected; the horns or lobes were covered with cilia for swimming and some of them bore suckers for use when the larvae should seek the firm underlying floor of the sea. During the short but critical period of attachment, the larval tissues were reorganized as completely as those of a pupal insect within a cocoon, the infant shape disappeared and in its place the five-rayed body of the adult was formed. Now as we find them, these new-made starfish use their tube feet competently, creeping over the rocks, righting their bodies if by mischance they are overturned, even, we may suppose, finding and devouring minute food animals in true starfish fashion.

The northern starfish lives in almost every low tide pool or waits out the tidal interval in wet moss or in the dripping coolness of a rock overhang. On a very low tide, when the departure of the sea is brief, these stars strew their variously hued forms over the moss like so many blossoms—pink, blue, purple, peach, or beige. Here and there is a gray or orange starfish on which the spines stand out conspicuously in a pattern of white dots. Its arms are rounder and firmer than those of the northern star and the round stony plate on its upper surface is usually a bright orange instead of pale yellow as in the northern species. This starfish is common south of Cape Cod and only a few individuals stray farther north. Still a third species inhabits these low-tide rocks—the blood-red starfish, Henricia, whose kind not only lives at these margins of the sea but goes down to lightless sea bottoms near the edge of the continental shelf. It is always an inhabitant of cool waters and south of Cape Cod must go offshore to find the temperatures it requires. But its dispersal is not, as one might suppose, by the larval stages, for unlike most other starfish it produces no swimming young; instead, the mother holds the eggs and the young that develop from them in a pouch formed by her arms as she assumes a humped position. Thus she broods them until they have become fully developed little starfish.

The Jonah crabs use the resilient cushion of moss as a hiding place to wait for the return of the tide or the coming of darkness. I remember a moss-carpeted ledge standing out from a rock wall, jutting out over sea depths where Laminaria rolled in the tide. The sea had only recently dropped below this ledge; its return was imminent and in fact was promised by every glassy swell that surged smoothly to its edge, then fell away. The moss was saturated, holding the water as faithfully as a sponge. Down within the deep pile of that carpet I caught a glimpse of a bright rosy color. At first I took it to be a growth of one of the encrusting corallines, but when I parted the fronds I was startled by abrupt movement as a large crab shifted its position and lapsed again into passive waiting. Only after search deep in the moss did I find several of the crabs, waiting out the brief interval of low tide and reasonably secure from detection by the gulls.

The seeming passivity of these northern crabs must be related to their need to escape the gulls—probably their most persistent enemies. By day one always has to search for the crabs. If not hidden deeply within the seaweeds, they may be wedged in the farthest recess afforded by an overhanging rock, secure there, in dim coolness, gently waving their antennae and waiting for the return of the sea. In darkness, however, the big crabs possess the shore. One night when the tide was ebbing I went down to the low-tide world to return a large starfish I had taken on the morning tide. The starfish was at home at the lowest level of these tides of the August moon, and to that level it must be returned. I took a flashlight and made my way down over the slippery rockweeds. It was an eerie world; ledges curtained with weed and boulders that by day were familiar landmarks seemed to loom larger than I remembered and to have assumed unfamiliar shapes, every projecting mass thrown into bold relief by the shadows. Everywhere I looked, directly in the beam of my flashlight or obliquely in the half-illuminated gloom, crabs were scuttling about. Boldly and possessively they inhabited the weed-shrouded rocks. All the grotesqueness of their form accentuated, they seemed to have transformed this once familiar place into a goblin world.

In some places, the moss is attached, not to the underlying rock, but to the next lower layer of life, a community of horse mussels. These large mollusks inhabit heavy, bulging shells, the smaller ends of which bristle with coarse yellow hairs that grow as excrescences from the epidermis. The horse mussels themselves are the basis of a whole community of animals that would find life on these wave-swept rocks impossible except for the presence and activities of the mollusks. The mussels have bound their shells to the underlying rock by an almost inextricable tangle of golden-hued byssus threads. These are the product of glands in the long slender foot, the threads being "spun" from a curious milky secretion that solidifies on contact with sea water. The threads possess a texture that is a remarkable combination of toughness, strength, softness, and elasticity; extending out in all directions they enable the mussels to hold their position not only against the thrust of incoming waves but also against the drag of the backwash, which in a heavy surf is tremendous.

Over the years that the mussels have been growing here, particles of muddy debris have settled under their shells and around the anchor lines of the byssus threads. This has created still another area for life, a sort of understory inhabited by a variety of animals including worms, crustaceans, echinoderms, and numerous mollusks, as well as the baby mussels of an oncoming generation—these as yet so small and transparent that the forms of their infant bodies show through newly formed shells.

Certain animals almost invariably live among the horse mussels. Brittle stars insinuate their thin bodies among the threads and under the shells of the mussels, gliding with serpentine motions of the long slender arms. The scale worms always live here, too, and down in the lower layers of this strange community of animals starfish may live below the scale worms and brittle stars, and sea urchins below the starfish, and sea cucumbers below the urchins.

Of the echinoderms that live here, few are the largest individuals of their species. The blanket of horse mussels seems to be a shelter for young, growing animals, and indeed the full-grown starfish and urchins could hardly be accommodated there. In the waterless intervals of the low tide, the cucumbers draw themselves into little football-shaped ovals scarcely more than an inch long; returned to the water and fully relaxed, they extend their bodies to a length of five or six inches and unfurl a crown of tentacles. The cucumbers are detritus feeders, and explore the surrounding muddy debris with their soft tentacles, which periodically they pull back and draw across their mouths, as a child would lick his fingers.

In pockets deep in the moss under layers of mussels, a long, slender little fish of the blenny tribe, the rock eel, waits for the return of the tide, coiled in its water-filled refuge with several of its kind. Disturbed by an intruder, all thrash the water violently, squirming with eel-like undulations to escape.

Where the big mussels grow more sparsely, in the seaward suburbs of this mussel city, the moss carpet, too, becomes a little thinner; but still the underlying rock seldom is exposed. The green crumb-of-bread sponge, which at higher levels seeks the shelter of rock overhangs and tide pools, here seems able to face the direct force of the sea and forms soft, thick mats of pale green, dotted with the cones and craters typical of this species. And here and there patches of another color show amid the thinning moss—dull rose or a gleaming, reddish brown of satin finish—an intimation of what lies at lower levels.

During much of the year the spring tides drop down into the band of Irish moss but go no lower, returning then toward the land. But in certain months, depending on the changing positions of sun and moon and earth, even the spring tides gain in amplitude, and their surge of water ebbs farther into the sea even as it rises higher against the land. Always, the autumn tides move strongly, and as the hunter's moon waxes and grows round, there come days and nights when the flood tides leap at the smooth rim of granite and send up their lace-edged wavelets to touch the roots of the bayberry; on their ebbs, with sun and moon combining to draw them back to the sea, they fall away from ledges not revealed since the April moon shone upon their dark shapes. Then they expose the sea's enameled floor—the rose of encrusting corallines, the green of sea urchins, the shining amber of the oarweeds.

At such a time of great tides I go down to that threshold of the sea world to which land creatures are admitted rarely in the cycle of the year. There I have known dark caves where tiny sea flowers bloom and masses of soft coral endure the transient withdrawal of the water. In these caves and in the wet gloom of deep crevices in the rocks I have found myself in the world of the sea anemones—creatures that spread a creamyhued crown of tentacles above the shining brown columns of their bodies, like handsome chrysanthemums blooming in little pools held in depressions or on bottoms just below the tide line.

Where they are exposed by this extreme ebbing of the water, their appearance is so changed that they seem not meant for even this brief experience of land life. Wherever the contours of this uneven sea floor provide some shelter I have found their exposed colonies—dozens or scores of anemones crowded together, their translucent bodies touching, side against side. The anemones that cling to horizontal surfaces respond to the withdrawal of water by pulling all their tissues down into a flattened, conical mass of firm consistency. The crown of feather-soft tentacles is retracted and tucked within, with no suggestion of the beauty that resides in an expanded anemone. Those that grow on vertical rocks hang down limply, extended into curious, hourglass shapes, all their tissues flaccid in the unaccustomed withdrawal of water. They do not lack the ability to contract, for when they are touched they promptly begin to shorten the column, drawing it up into more normal proportions. These anemones, deserted by the sea, are bizarre objects rather than things of beauty, and indeed bear only the most remote resemblance to the anemones blooming under water just offshore, all their tentacles expanded in the search for food. As small water creatures come in contact with the tentacles of these expanded anemones, they receive a deadly discharge. Each of the thousand or more tentacles bears thousands of coiled darts embedded in its substance, each with a minute spine protruding. The spine may act as a trigger to set off the explosion, or perhaps the very nearness of prey acts as a sort of chemical trigger, causing the dart to explode with great violence, impaling or entangling its victim and injecting a poison.

Like the anemones, the soft coral hangs its thimble-sized colonies on the under side of ledges. Limp and dripping at low tide, they suggest nothing of the life and beauty to which the returning water restores them. Then from all the myriad pores of the surface of the colony, the tentacles of little tubular animals appear and the polyps thrust themselves out into the tide, seizing each for itself the minute shrimps and copepods and multiformed larvae brought by the water.

The soft coral, or sea finger, secretes no limy cups as the distantly related stony, or reef, corals do, but forms colonies in which many animals live embedded in a tough matrix strengthened with spicules of lime. Minute though the spicules are, they become geologically important where, in tropical reefs, the soft corals, or Alcyonaria, mingle with the true corals. With the death and dissolution of the soft tissues, the hard spicules become minute building stones, entering into the composition of the reef. Alcyonarians grow in lush profusion and variety on the coral reefs and flats of the Indian Ocean, for these soft corals are predominantly creatures of the tropics. A few, however, venture into polar waters. One very large species, tall as a tall man and branched like a tree, lives on the fishing banks off Nova Scotia and New England. Most of the group live in deep waters; for the most part the intertidal rocks are inhospitable to them and only an occasional low-lying ledge, rarely and briefly exposed on the low spring tides, bears their colonies on dark and hidden surfaces.

In seams and crevices of rock, in little water-filled pools, or on rock walls briefly exposed by the tide's low ebbing, colonies of the pink-hearted hydroid Tubularia form gardens of beauty. Where the water still covers them the flowerlike animals sway gracefully at the ends of long stalks, their tentacles reaching out to capture small animals of the plankton. Perhaps it is where they are permanently submerged, however, that they reach their fullest development. I have seen them coating wharf pilings, floats, and submerged ropes and cables so thickly that not a trace of the substratum could be seen, their growth giving the illusion of thousands of blossoms, each as large as the tip of my little finger.

Below the last clumps of Irish moss, a new kind of sea bottom is exposed. The transition is abrupt. As though a line had been drawn, suddenly there is no more moss, and one steps from the yielding brown cushion onto a surface that seemingly is of stone. Except that the color is wrong, the effect is almost that of a volcanic slope—there is the same barren nakedness. Yet this is not rock that we see. The underlying rock is coated on every surface, vertical or horizontal, exposed or hidden, with a crust of coralline algae, so that it wears a rich old-rose color. So intimate is the union that the plant seems part of the rock. Here the periwinkles wear little patches of pink on their shells, all the rock caverns and fissures are lined with the same color, and the rock bottom that slants away into green water carries down the rose hue as far as the eye can follow.

The coralline algae are plants of unusual fascination. They belong to the group of red seaweeds, most of which live in the deeper coastal waters, for the chemical nature of their pigments usually requires the protection of a screen of water between their tissues and the sun. The corallines, however, are extraordinary in their ability to withstand direct sunlight. They are able to incorporate carbonate of lime into their tissues so that they have become hardened. Most species form encrusting patches on rocks, shells, and other firm surfaces. The crust may be thin and smooth, suggesting a coat of enamel paint; or it may be thick and roughened by small nodules and protuberances. In the tropics the corallines often enter importantly into the composition of the coral reefs, helping to cement the branching structures built by the coral animals into a solid reef. Here and there in the East Indies they cover the tidal flats as far as eye can see with their delicately hued crusts, and many of the "coral reefs" of the Indian Ocean contain no coral but are built largely of these plants. About the coasts of Spitsbergen, where under the dimly lit waters of the north the great forests of the brown algae grow, there are also vast calcareous banks, stretching mile after mile, formed by the coralline algae. Being able to live not only in tropical warmth but where water temperatures seldom rise above the freezing point, these plants flourish all the way from Arctic to Antarctic seas.

Where these same corallines paint a rose-colored band on the rocks of the Maine coast, as though to mark the low water line of the lowest spring tides, visible animal life is scarce. But although little else lives openly in this zone, thousands of sea urchins do. Instead of hiding in crevices or under rocks as they do at the higher levels, they live fully exposed on the flat or gently shelving rock faces. Groups of a score or half a hundred individuals lie together on the coralline-coated rocks, forming patches of pure green on the rose background. I have seen such herds of urchins lying on rocks that were being washed by a heavy surf, but apparently all the little anchors formed by their tube feet held securely. Though the waves broke heavily and poured back in a turbulent rush of waters, there the urchins remained undisturbed. Perhaps the strong tendency to hide and to wedge themselves into crevices and under boulders, as displayed by urchins in tide pools or up in the rockweed zone, is not so much an attempt to avoid the power of the surf as a means of escaping the eager eyes of the gulls, who hunt them relentlessly on every low tide. This coralline zone where the urchins live so openly is covered almost constantly with a protective layer of water; probably not more than a dozen daytime tides in the entire year fall to this level. At all other times, the depth of water over the urchins prevents the gulls from reaching them, for although a gull can make shallow plunges under water, it cannot dive as a tern does, and probably cannot reach a bottom deeper than the length of its own body.

The lives of many of these creatures of the low-tide rocks are bound together by interlacing ties, in the relation of predator to prey, or in the relation of species that compete for space or food. Over all these the sea itself exercises a directing and regulating force.

The urchins seek sanctuary from the gulls at this low level of the spring tides, but in themselves stand in the relation of dangerous predators to other animals. Where they advance into the Irish moss zone, hiding in deep crevices and sheltering under rock overhangs, they devour numbers of periwinkles, and even attack barnacles and mussels. The number of urchins at any particular level of shore has a strong regulating effect on the populations of their prey. The starfish and a voracious snail, the common whelk, like the urchins, have their centers of population in deep water offshore and make predatory excursions of varying duration into the intertidal zone.

The position of the prey animals—the mussels, barnacles, and periwinkles—on sheltered shores has become difficult. They are hardy and adaptable, able to live at any level of the tide. Yet on such shores the rockweeds have crowded them out of the upper two-thirds of the shore, except for scattered individuals. At and just below the low-tide line are the hungry predators, so all that remains for these animals is the level near the low-water line of the neap tides. On protected coasts it is here that the barnacles and mussels assemble in their millions to spread their cover of white and blue over the rocks, and the legions of the common periwinkle gather.

But the sea, with its tempering and modifying effect, can alter the pattern. Whelks, starfish, and urchins are creatures of cold water. Where the offshore waters are cold and deep and the tidal flow is drawn from these icy reservoirs, the predators can range up into the intertidal zone, decimating the numbers of their prey. But when there is a layer of warm surface water the predators are confined to the cold deep levels. As they retreat seaward, the legions of their prey follow down in their wake, descending as far as they may into the world of the low spring tides.

Tide pools contain mysterious worlds within their depths, where all the beauty of the sea is subtly suggested and portrayed in miniature. Some of the pools occupy deep crevices or fissures; at their seaward ends these crevices disappear under water, but toward the land they run back slantingly into the cliffs and their walls rise higher, casting deep shadows over the water within them. Other pools are contained in rocky basins with a high rim on the seaward side to hold back the water when the last of the ebb drains away. Seaweeds line their walls. Sponges, hydroids, anemones, sea slugs, mussels, and starfish live in water that is calm for hours at a time, while just beyond the protecting rim the surf may be pounding.

The pools have many moods. At night they hold the stars and reflect the light of the Milky Way as it flows across the sky above them. Other, living stars come in from the sea: the shining emeralds of tiny phosphorescent diatoms—the glowing eyes of small fishes that swim at the surface of the dark water, their bodies slender as matchsticks, moving almost upright with little snouts uplifted—the elusive moonbeam flashes of comb jellies that have come in with a rising tide. Fishes and comb jellies hunt the black recesses of the rock basins, but like the tides they come and go, having no part in the permanent life of the pools.

By day there are other moods. Some of the most beautiful pools lie high on the shore. Their beauty is the beauty of simple elements—color and form and reflection. I know one that is only a few inches deep, yet it holds all the depth of the sky within it, capturing and confining the reflected blue of far distances. The pool is outlined by a band of bright green, a growth of one of the seaweeds called Enteromorpha. The fronds of the weed are shaped like simple tubes or straws. On the land side a wall of gray rock rises above the surface to the height of a man, and reflected, descends its own depth into the water. Beyond and below the reflected cliff are those far reaches of the sky. When the light and one's mood are right, one can look down into the blue so far that one would hesitate to set foot in so bottomless a pool. Clouds drift across it and wind ripples scud over its surface, but little else moves there, and the pool belongs to the rock and the plants and the sky.

In another high pool nearby, the green tube-weed rises from all of the floor. By some magic the pool transcends its realities of rock and water and plants, and out of these elements creates the illusion of another world. Looking into the pool, one sees no water but instead a pleasant landscape of hills and valleys with scattered forests. Yet the illusion is not so much that of an actual landscape as of a painting of one; like the strokes of a skillful artist's brush, the individual fronds of the algae do not literally portray trees, they merely suggest them. But the artistry of the pool, as of the painter, creates the image and the impression.

Little or no animal life is visible in any of these high pools—perhaps a few periwinkles and a scattering of little amber isopods. Conditions are difficult in all pools high on the shore because of the prolonged absence of the sea. The temperature of the water may rise many degrees, reflecting the heat of the day. The water freshens under heavy rains or becomes more salty under a hot sun. It varies between acid and alkaline in a short time through the chemical activity of the plants. Lower on the shore the pools provide far more stable conditions, and both plants and animals are able to live at higher levels than they could on open rock. The tide pools, then, have the effect of moving the life zones a little higher on the shore. Yet they, too, are affected by the duration of the sea's absence, and the inhabitants of a high pool are very different from those of a low-level pool that is separated from the sea only at long intervals and then briefly.

The highest of the pools scarcely belong to the sea at all; they hold the rains and receive only an occasional influx of sea water from storm surf or very high tides. But the gulls fly up from their hunting at the sea's edge, bringing a sea urchin or a crab or a mussel to drop on the rocks, in this way shattering the hard shelly covering and exposing the soft parts within. Bits of urchin tests or crab claws or mussel shells find their way into the pools, and as they disintegrate their limy substance enters into the chemistry of the water, which then becomes alkaline. A little one-celled plant called Sphaerella finds this a favorable climate for growth—a minute, globular bit of life almost invisible as an individual, but in its millions turning the waters of these high pools red as blood. Apparently the alkalinity is a necessary condition; other pools, outwardly similar except for the chance circumstance that they contain no shells, have none of the tiny crimson balls.

Even the smallest pools, filling depressions no larger than a teacup, have some life. Often it is a thin patch of scores of the little seashore insect, Anurida maritima—"the wingless one who goes to sea." These small insects run on the surface film when the water is undisturbed, crossing easily from one shore of a pool to another. Even the slightest rippling causes them to drift helplessly, however, so that scores or hundreds of them come together by chance, becoming conspicuous only as they form thin, leaflike patches on the water. A single Anurida is small as a gnat. Under a lens, it seems to be clothed in blue-gray velvet through which many bristles or hairs protrude. The bristles hold a film of air about the body of the insect when it enters the water, and so it need not return to the upper shore when the tide rises. Wrapped in its glistening air blanket, dry and provided with air for breathing, it waits in cracks and crevices until the tide ebbs again. Then it emerges to roam over the rocks, searching for the bodies of fish and crabs and the dead mollusks and barnacles that provide its food, for it is one of the scavengers that play a part in the economy of the sea, keeping the organic materials in circulation.

And often I find the pools of the upper third of the shore lined with a brown velvety coating. My fingers, exploring, are able to peel it off the rocks in thin smooth-surfaced sheets like parchment. It is one of the brown seaweeds called Ralfsia; it appears on the rocks in small, lichen-like growths or, as here, spreading its thin crust over extensive areas. Wherever it grows its presence changes the nature of a pool, for it provides the shelter that many small creatures seek so urgently. Those small enough to creep in under it—to inhabit the dark pockets of space between the encrusting weed and the rock—have found security against being washed away by the surf. Looking at these pools with their velvet lining, one would say there is little life here—only a sprinkling of periwinkles browsing, their shells rocking gently as they scrape at the surface of the brown crust, or perhaps a few barnacles with their cones protruding through the sheet of plant tissue, opening their doors to sweep the water for food. But whenever I have brought a sample of this brown seaweed to my microscope, I have found it teeming with life. Always there have been many cylindrical tubes, needle-fine, built of a muddy substance. The architect of each is a small worm whose body is formed of a series of eleven infinitely small rings or segments, like eleven counters in a game of checkers, piled one above another. From its head arises a structure that makes this otherwise drab worm beautiful—a fanlike crown or plume composed of the finest feathery filaments. The filaments absorb oxygen and also serve to ensnare small food organisms when thrust out of the tube. And always, among this microfauna of the Ralfsia crust, there have been little fork-tailed crustaceans with glittering eyes the color of rubies. Other crustaceans called ostracods are enclosed in flattened, peach-colored shells fashioned of two parts, like a box with its lid; from the shell long appendages may be thrust out to row the creatures through the water. But most numerous of all are the minute worms hurrying across the crust—segmented bristle worms of many species and smooth-bodied, serpent-like ribbon worms or nemerteans, their appearance and rapid movements betraying their predatory errands.

A pool need not be large to hold beauty within pellucid depths. I remember one that occupied the shallowest of depressions; as I lay outstretched on the rocks beside it I could easily touch its far shore. This miniature pool was about midway between the tide lines, and for all I could see it was inhabited by only two kinds of life. Its floor was paved with mussels. Their shells were a soft color, the misty blue of distant mountain ranges, and their presence lent an illusion of depth. The water in which they lived was so clear as to be invisible to my eyes; I could detect the interface between air and water only by the sense of coldness on my fingertips. The crystal water was filled with sunshine—an infusion and distillation of light that reached down and surrounded each of these small but resplendent shellfish with its glowing radiance.

The mussels provided a place of attachment for the only other visible life of the pool. Fine as the finest threads, the basal stems of colonies of hydroids traced their almost invisible lines across the mussel shells. The hydroids belonged to the group called Sertularia, in which each individual of the colony and all the supporting and connecting branches are enclosed within transparent sheaths, like a tree in winter wearing a sheath of ice. From the basal stems erect branches arose, each branch the bearer of a double row of crystal cups within which the tiny beings of the colony dwelt. The whole was the very embodiment of beauty and fragility, and as I lay beside the pool and my lens brought the hydroids into clearer view they seemed to me to look like nothing so much as the finest cut glass—perhaps the individual segments of an intricately wrought chandelier. Each animal in its protective cup was something like a very small sea anemone—a little tubular being surmounted by a crown of tentacles. The central cavity of each communicated with a cavity that ran the length of the branch that bore it, and this in turn with the cavities of larger branches and with those of the main stem, so that the feeding activities of each animal contributed to the nourishment of the whole colony.

On what, I wondered, were these Sertularians feeding? From their very abundance I knew that whatever creatures served them as food must be infinitely more numerous than the carnivorous hydroids themselves. Yet I could see nothing. Obviously their food would be minute, for each of the feeders was of threadlike diameter and its tentacles were like the finest gossamer. Somewhere in the crystal clarity of the pool my eye—or so it seemed—could detect a fine mist of infinitely small particles, like dust motes in a ray of sunshine. Then as I looked more closely the motes had disappeared and there seemed to be once more only that perfect clarity, and the sense that there had been an optical illusion. Yet I knew it was only the human imperfection of my vision that prevented me from seeing those microscopic hordes that were the prey of the groping, searching tentacles I could barely see. Even more than the visible life, that which was unseen came to dominate my thoughts, and finally the invisible throng seemed to me the most powerful beings in the pool. Both the hydroids and the mussels were utterly dependent on this invisible flotsam of the tide streams, the mussels as passive strainers of the plant plankton, the hydroids as active predators seizing and ensnaring the minute water fleas and copepods and worms. But should the plankton become less abundant, should the incoming tide streams somehow become drained of this life, then the pool would become a pool of death, both for the mussels in their shells blue as mountains and for the crystal colonies of the hydroids.

Some of the most beautiful pools of the shore are not exposed to the view of the casual passer-by. They must be searched for—perhaps in low-lying basins hidden by great rocks that seem to be heaped in disorder and confusion, perhaps in darkened recesses under a projecting ledge, perhaps behind a thick curtain of concealing weeds.

I know such a hidden pool. It lies in a sea cave, at low tide filling perhaps the lower third of its chamber. As the flooding tide returns the pool grows, swelling in volume until all the cave is water-filled and the cave and the rocks that form and contain it are drowned beneath the fullness of the tide. When the tide is low, however, the cave may be approached from the landward side. Massive rocks form its floor and walls and roof. They are penetrated by only a few openings—two near the floor on the sea side and one high on the landward wall. Here one may lie on the rocky threshold and peer through the low entrance into the cave and down into its pool. The cave is not really dark; indeed on a bright day it glows with a cool green light. The source of this soft radiance is the sunlight that enters through the openings low on the floor of the pool, but only after its entrance into the pool does the light itself become transformed, invested with a living color of purest, palest green that is borrowed from the covering of sponge on the floor of the cave.

Through the same openings that admit the light, fish come in from the sea, explore the green hall, and depart again into the vaster waters beyond. Through those low portals the tides ebb and flow. Invisibly, they bring in minerals—the raw materials for the living chemistry of the plants and animals of the cave. They bring, invisibly again, the larvae of many sea creatures—drifting, drifting in their search for a resting place. Some may remain and settle here; others will go out on the next tide.

Looking down into the small world confined within the walls of the cave, one feels the rhythms of the greater sea world beyond. The waters of the pool are never still. Their level changes not only gradually with the rise and fall of the tide, but also abruptly with the pulse of the surf. As the backwash of a wave draws it seaward, the water falls away rapidly; then with a sudden reversal the inrushing water foams and surges upward almost to one's face.

On the outward movement one can look down and see the floor, its details revealed more clearly in the shallowing water. The green crumb-of-bread sponge covers much of the bottom of the pool, forming a thick-piled carpet built of tough little feltlike fibers laced together with glassy, double-pointed needles of silica—the spicules or skeletal supports of the sponge. The green color of the carpet is the pure color of chlorophyll, this plant pigment being confined within the cells of an alga that are scattered through the tissues of the animal host. The sponge clings closely to the rock, by the very smoothness and flatness of its growth testifying to the streamlining force of heavy surf. In quiet waters the same species sends up many projecting cones; here these would give the turbulent waters a surface to grip and tear.

Interrupting the green carpet are patches of other colors, one a deep, mustard yellow, probably a growth of the sulphur sponge. In the fleeting moment when most of the water has drained away, one has glimpses of a rich orchid color in the deepest part of the cave—the color of the encrusting coralline algae.

Sponges and corallines together form a background for the larger tide-pool animals. In the quiet of ebb tide there is little or no visible movement even among the predatory starfish that cling to the walls like ornamental fixtures painted orange or rose or purple. A group of large anemones lives on the wall of the cave, their apricot color vivid against the green sponge. Today all the anemones may be attached on the north wall of the pool, seemingly immobile and immovable; on the next spring tides when I visit the pool again some of them may have shifted over to the west wall and there taken up their station, again seemingly immovable.

There is abundant promise that the anemone colony is a thriving one and will be maintained. On the walls and ceiling of the cave are scores of baby anemones—little glistening mounds of soft tissue, a pale, translucent brown. But the real nursery of the colony seems to be in a sort of antechamber opening into the central cave. There a roughly cylindrical space no more than a foot across is enclosed by high perpendicular rock walls to which hundreds of baby anemones cling.

On the roof of the cave is written a starkly simple statement of the force of the surf. Waves entering a confined space always concentrate all their tremendous force for a driving, upward leap in this manner the roofs of caves are gradually battered away. The open portal in which I lie saves the ceiling of this cave from receiving the full force of such upward-leaping waves; nevertheless, the creatures that live there are exclusively a heavy-surf fauna. It is a simple black and white mosaic— the black of mussel shells, on which the white cones of barnacles are growing. For some reason the barnacles, skilled colonizers of surf-swept rocks though they be, seem to have been unable to get a foothold directly on the roof of the cave. Yet the mussels have done so. I do not know how this happened but I can guess. I can imagine the young mussels creeping in over the damp rock while the tide is out, spinning their silk threads that bind them securely, anchoring them against the returning waters. And then in time, perhaps, the growing colony of mussels gave the infant barnacles a foothold more tenable than the smooth rock, so that they were able to cement themselves to the mussel shells. However it came about, that is the way we find them now.

As I lie and look into the pool there are moments of relative quiet, in the intervals when one wave has receded and the next has not yet entered. Then I can hear the small sounds: the sound of water dripping from the mussels on the ceiling or of water dripping from seaweeds that line the walls—small, silver splashes losing themselves in the vastness of the pool and in the confused, murmurous whisperings that emanate from the pool itself—the pool that is never quite still.

Then as my fingers explore among the dark red thongs of the dulse and push away the fronds of the Irish moss that cover the walls beneath me, I begin to find creatures of such extreme delicacy that I wonder how they can exist in this cave when the brute force of storm surf is unleashed within its confined space.

Adhering to the rock walls are thin crusts of one of the bryozoans, a form in which hundreds of minute, flask-shaped cells of a brittle structure, fragile as glass, lie one against another in regular rows to form a continuous crust. The color is a pale apricot; the whole seems an ephemeral creation that would crumble away at a touch, as hoarfrost before the sun.

A tiny spiderlike creature with long and slender legs runs about over the crust. For some reason that may have to do with its food, it is the same apricot color as the bryozoan carpet beneath it; the sea spider, too, seems the embodiment of fragility.

Another bryozoan of coarser, upright growth, Flustrella, sends up little club-shaped projections from a basal mat. Again, the lime-impregnated clubs seem brittle and glassy. Over and among them, innumerable little roundworms crawl with serpentine motion, slender as threads. Baby mussels creep in their tentative exploration of a world so new to them they have not yet found a place to anchor themselves by slender silken lines.

Exploring with my lens, I find many very small snails in the fronds of seaweed. One of them has obviously not been long in the world, for its pure white shell has formed only the first turn of the spiral that will turn many times upon itself in growth from infancy to maturity. Another, no larger, is nevertheless older. Its shining amber shell is coiled like a French horn and, as I watch, the tiny creature within thrusts out a bovine head and seems to be regarding its surroundings with two black eyes, small as the smallest pinpoints.

But seemingly most fragile of all are the little calcareous sponges that here and there exist among the seaweeds. They form masses of minute, upthrust tubes of vase-like form, none more than half an inch high. The wall of each is a mesh of fine threads—a web of starched lace made to fairy scale.

I could have crushed any of these fragile structures between my fingers—yet somehow they find it possible to exist here, amid the surging thunder of the surf that must fill this cave as the sea comes in. Perhaps the seaweeds are the key to the mystery, their resilient fronds a sufficient cushion for all the minute and delicate beings they contain.

But it is the sponges that give to the cave and its pool their special quality—the sense of a continuing flow of time. For each day that I visit the pool on the lowest tides of the summer they seem unchanged—the same in July, the same in August, the same in September. And they are the same this year as last, and presumably as they will be a hundred or a thousand summers hence.

Simple in structure, little different from the first sponges that spread their mats on ancient rocks and drew their food from a primordial sea, the sponges bridge the eons of time. The green sponge that carpets the floor of this cave grew in other pools before this shore was formed; it was old when the first creatures came out of the sea in those ancient eras of the Paleozoic, 300 million years ago; it existed even in the dim past before the first fossil record, for the hard little spicules—all that remains when the living tissue is gone—are found in the first fossil-bearing rocks, those of the Cambrian period.

So, in the hidden chamber of that pool, time echoes down the long ages to a present that is but a moment.

As I watched, a fish swam in, a shadow in the green light, entering the pool by one of the openings low on its seaward wall. Compared with the ancient sponges, the fish was almost a symbol of modernity, its fishlike ancestry traceable only half as far into the past. And I, in whose eyes the images of the two were beheld as though they were contemporaries, was a mere newcomer whose ancestors had inhabited the earth so briefly that my presence was almost anachronistic.

As I lay at the threshold of the cave thinking those thoughts, the surge of waters rose and flooded across the rock on which I rested. The tide was rising.

IV
THE RIM OF SAND

ON THE SANDS of the sea's edge, especially where they are broad and bordered by unbroken lines of wind-built dunes, there is a sense of antiquity that is missing from the young rock coast of New England. It is in part a sense of the unhurried deliberation of earth processes that move with infinite leisure, with all eternity at their disposal. For unlike that sudden coming in of the sea to flood the valleys and surge against the mountain crests of the drowned lands of New England, the sea and the land lie here in a relation established gradually, over millions of years.

During those long ages of geologic time, the sea has ebbed and flowed over the great Atlantic coastal plain. It has crept toward the distant Appalachians, paused for a time, then slowly receded, sometimes far into its basin; and on each such advance it has rained down its sediments and left the fossils of its creatures over that vast and level plain. And so the particular place of its stand today is of little moment in the history of the earth or in the nature of the beach—a hundred feet higher, or a hundred feet lower, the seas would still rise and fall unhurried over shining flats of sand, as they do today.

And the materials of the beach are themselves steeped in antiquity. Sand is a substance that is beautiful, mysterious, and infinitely variable; each grain on a beach is the result of processes that go back into the shadowy beginnings of life, or of the earth itself.

The bulk of seashore sand is derived from the weathering and decay of rocks, transported from their place of origin to the sea by the rains and the rivers. In the unhurried processes of erosion, in the freighting seaward, ±n the interruptions and resumptions of that journey, the minerals have suffered various fates—some have been dropped, some have worn out and vanished. In the mountains the slow decay and disintegration of the rocks proceed, and the stream of sediments grows—suddenly and dramatically by rockslides—slowly, inexorably, by the wearing of rock by water. All begin their passage toward the sea. Some disappear through the solvent action of water or by grinding attrition in the rapids of a river's bed. Some are dropped on the riverbank by flood waters, there to lie for a hundred, a thousand years, to become locked in the sediments of the plain and wait another million years or so, during which, perhaps, the sea comes in and then returns to its basin. Then at last they are released by the persistent work of erosion's tools-wind, rain, and frost—to resume the journey to the sea. Once brought to salt water, a fresh rearranging, sorting, and transport begin. Light minerals, like flakes of mica, are carried away almost at once; heavy ones like the black sands of ilmenite and rutile are picked up by the violence of storm waves and thrown on the upper beach.

No individual sand grain remains long in any one place. The smaller it is, the more it is subject to long transport—the larger grains by water, the smaller by wind. An average grain of sand is only two and one half times the weight of an equal volume of water, but more than two thousand times as heavy as air, so only the smaller grains are available for transport by wind. But despite the constant working over of the sands by wind and water, a beach shows little visible change from day to day, for as one grain is carried away, another is usually brought to take its place.

The greater part of most beach sand consists of quartz, the most abundant of all minerals, found in almost every type of rock. But many other minerals occur among its crystal grains, and one small sample of sand might contain fragments of a dozen or more. Through the sorting action of wind, water, and gravity, fragments of darker, heavier minerals may form patches overlying the pale quartz. So there may be a curious purple shading over the sand, shifting with the wind, piling up in little ridges of deeper color like the ripple marks of waves—a concentration of almost pure garnet. Or there may be patches of dark green—sands formed of glauconite, a product of the sea's chemistry and the interaction of the living and the non-living. Glauconite is a form of iron silicate that contains potassium; it has occurred in the deposits of all geologic ages. According to one theory, it is forming now in warm shallow areas of the sea's floor, where the shells of minute creatures called foraminifera are accumulating and disintegrating on muddy sea bottoms. On many Hawaiian beaches, the somber darkness of the earth's interior is reflected in sand grains of olivine derived from black basaltic lavas. And drifts of the "black sands" of rutile and ilmenite and other heavy minerals darken the beaches of Georgia's St. Simons and Sapelo Islands, clearly separated from the lighter quartz.

In some parts of the world the sands represent the remains of plants that in life had lime-hardened tissues, or fragments of the calcareous shells of sea creatures. Here and there on the coast of Scotland, for example, are beaches composed of glistening white "nullipore sands"—the shattered and sea-ground remains of coralline algae growing on the bottom offshore. On the coast of Galway in Ireland the dunes are built of sands composed of tiny perforated globes of calcium carbonate—the shells of foraminifera that once floated in the sea. The animals were mortal but the shells they built have endured. They drifted to the floor of the sea and became compacted into sediment. Later the sediments were uplifted to form cliffs, which were eroded and returned once more to the sea. The shells of foraminifera appear also in the sands of southern Florida and the Keys, along with coral debris and the shells of mollusks, shattered, ground, and polished by the waves.

From Eastport to Key West, the sands of the American Atlantic coast, by their changing nature, reveal a varied origin. Toward the northern part of the coast mineral sands predominate, for the waves are still sorting and rearranging and carrying from place to place the fragments of rock that the glaciers brought down from the north, thousands of years ago. Every grain of sand on a New England beach has a long and eventful history. Before it was sand, it was rock—splintered by the chisels of the frost, crushed under advancing glaciers and carried forward with the ice in its slow advance, then ground and polished in the mill of the surf. And long ages before the advance of the ice, some of the rock had come up into the light of the sun from the black interior of the earth by ways unseen and for the most part unknown, made fluid by subterranean fires and rising along deep pipes and fissures. Now in this particular moment of its history, it belongs to the sea's edge—swept up and down the beaches with the tides or drifted alongshore with the currents, continuously sifted and sorted, packed down, washed out, or set adrift again, as always and endlessly the waves work over the sands.

On Long Island, where much glacial material has accumulated, the sands contain quantities of pink and red garnet and black tourmaline, along with many grains of magnetite. In New Jersey, where the coastal plain deposits of the south first appear, there is less magnetic material and less garnet. Smoky quartz predominates at Barnegat, glauconite at Monmouth Beach, and heavy minerals at Cape May. Here and there beryl occurs where molten magma has brought up deeply buried material of the ancient earth to crystallize near the surface.

North of Virginia, less than half of one per cent of the sands are of calcium carbonate; southward, about 5 per cent. In North Carolina the abundance of calcareous or shell sand suddenly increases, although quartz sand still forms the bulk of the beach materials. Between Capes Hatteras and Lookout as much as 10 per cent of the beach sand is calcareous. And in North Carolina also there are odd local accumulations of special materials such as silicified wood—the same substance that is contained in the famous "singing sands" of the Island of Eigg in the Hebrides.

The mineral sands of Florida are not of local origin but have been derived from the weathering of rocks in the Piedmont and Appalachian highlands of Georgia and South Carolina. The fragments are carried to the sea on southward-moving streams and rivers. Beaches of the northern part of Florida's Gulf Coast are almost pure quartz, composed of crystal grains that have descended from the mountains to sea level, accumulating there in plains of snowlike whiteness. About Venice there is a special sparkle and glitter over the sands, where crystals of the mineral zircon are dusted over its surface like diamonds; and here and there is a sprinkling of the blue, glasslike grains of cyanite. On the east coast of Florida, quartz sands predominate for much of the long coast line (it is the hard-packing quartz grains that compose the famous beaches of Daytona) but toward the south, the crystal sands are mingled more and more with fragments of shells. Near Miami the beach sands are less than half quartz; about Cape Sable and in the Keys the sand is almost entirely derived from coral and shell and the remains of foraminifera. And all along the east coast of Florida, the beaches receive small contributions of volcanic matter, as bits of floating pumice that have drifted for thousands of miles in ocean currents are stranded on the shore to become sand.

Infinitely small though it is, something of its history may be revealed in the shape and texture of a grain of sand. Wind-transported sands tend to be better rounded than water-borne; furthermore, their surface shows a frosted effect from the abrasion of other grains carried in the blast of air. The same effect is seen on panes of glass near the sea, or on old bottles in the beach flotsam. Ancient sand grains, by their surface etchings, may give a clue to the climate of past ages. In European deposits of Pleistocene sand, the grains have frosted surfaces etched by the great winds blowing off the glaciers of the Ice Age.

We think of rock as a symbol of durability, yet even the hardest rock shatters and wears away when attacked by rain, frost or surf. But a grain of sand is almost indestructible. It is the ultimate product of the work of the waves—the minute, hard core of mineral that remains after years of grinding and polishing. The tiny grains of wet sand lie with little space between them, each holding a film of water about itself by capillary attraction. Because of this cushioning liquid film, there is little further wearing by attrition. Even the blows of heavy surf cannot cause one sand grain to rub against another.

In the intertidal zone, this minuscule world of the sand grains is also the world of inconceivably minute beings, which swim through the liquid film around a grain of sand as fish would swim through the ocean covering the sphere of the earth. Among this fauna and flora of the capillary water are single-celled animals and plants, water mites, shrimplike crustacea, insects, and the larvae of certain infinitely small worms—all living, dying, swimming, feeding, breathing, reproducing in a world so small that our human senses cannot grasp its scale, a world in which the micro-droplet of water separating one grain of sand from another is like a vast, dark sea.

Not all sands are inhabited by this "interstitial fauna." Those derived from the weathering of crystalline rocks are most abundantly populated. Shell or coral sand seldom if ever contains copepods and other microscopic life; perhaps this indicates that the grains of calcium carbonate create unfavorably alkaline conditions in the water around them.

On any beach the sum of all the little pools amid the sand grains represents the amount of water available to the animals of the sands during the low-tide interval. Sand of average fineness is able to contain almost its own volume of water, and so at low tide only the topmost layers dry out under a warm sun. Below it is damp and cool, for the contained water keeps the temperatures of the deeper sand practically constant. Even the salinity is fairly stable; only the most superficial layers are affected by rain falling on the beach or by streams of fresh water coursing across it.

Bearing on its surface only the wave-carved ripple marks, the fine traceries of sand grains dropped at last by the spent waves, and the scattered shells of long-dead mollusks, the beach has a lifeless look, as though not only uninhabited but indeed uninhabitable. In the sands almost all is hidden. The only clues to the inhabitants of most beaches are found in winding tracks, in slight movements disturbing the upper layers, or in barely protruding tubes and all but concealed openings leading down to hidden burrows.

The signs of living creatures are often visible, if not the animals themselves, in deep gullies that cut the beaches, parallel to the shore line, and hold at least a few inches of water from the fall of one tide until the return of the next. A little moving hill of sand may yield a moon snail intent on predatory errand. A V-shaped track may indicate the presence of a burrowing clam, a sea mouse, a heart urchin. A flat ribbonlike track may lead to a buried sand dollar or a starfish. And wherever protected flats of sand or sandy mud lie exposed between the tides, they are apt to be riddled with hundreds of holes, marked by the sign of the ghost shrimps within. Other flats may bristle with forests of protruding tubes, pencil thin and decorated weirdly with bits of shell or seaweed, an indication that legions of the plumed worm, Diopatra, live below. Or again there may be a wide area marked by the black conical mounds of the lugworm. Or here at the edge of the tide a chain of little parchment capsules, one end free and the other disappearing under the sand, shows that one of the large predatory whelks lies below, busy with the prolonged task of laying and protecting her eggs.

But almost always the essence of the lives—the finding of food, the hiding from enemies, the capturing of prey, the producing of young, all that makes up the living and dying and perpetuating of this sand-beach fauna—is concealed from the eyes of those who merely glance at the surface of the sands and declare them barren.

I remember a chill December morning on one of Florida's Ten Thousand Islands, with the sands wet from a recently fallen tide and the fresh, clean wind blowing handfuls of spindrift along the beach. For several hundred yards, where the shore ran in a long curve from the Gulf toward the shelter of the bay, there were peculiar markings on the dark wet sand just above the water's edge. The marks were arranged in groups, in each of which a series of thin spidery lines radiated out from a central spot, as though unsteadily traced there by a slender stick. At first no sign of any living animal was to be seen—

nothing to tell what creature had made these seemingly careless scribblings. After kneeling on the wet sand and looking at one after another of these strange insignia, I found that under each of the central spots lay the flat pentagonal disc of a serpent starfish. The marks on the sand were made by its long and slender arms, inscribing the record of its forward progress.

And then I remember wading on a June day over Bird Shoal, which lies off the town of Beaufort in North Carolina, where at low tide acres of sand bottom are covered only by a few inches of water. Near the shore I found two sharply defined grooves in the sand; my index finger could have measured their span. Between the grooves was a faint, irregular line. Step by step, I was led out across the flat by the tracks; finally, at the temporary end of the trail, I came upon a young horseshoe crab, heading seaward.

For most of the fauna of the sand beaches, the key to survival is to burrow into the wet sand, and to possess means of feeding, breathing and reproducing while lying below reach of the surf. And so the story of the sand is in part the story of small lives lived deep within it, finding in its dark, damp coolness a retreat from fish that come hunting with the tide and from birds that forage at the water's edge when the tide has fallen. Once below the surface layers, the burrower has found not only stable conditions but also a refuge where few enemies threaten. Those few are likely to reach down from above—perhaps a bird thrusting a long bill into the hole of a fiddler crab—a sting ray flapping along the bottom, plowing up the sand for buried mollusks—an octopus sliding an exploring tentacle down into a hole. Only an occasional enemy comes through the sand. The moon snail is a predator that makes a successful living in this difficult way. It is a blind creature with no use for eyes because it is forever groping through dark sands, hunting mollusks that live as much as a foot below the surface. Its smoothly rounded shell eases its descent into the sand as it digs with the immense foot. On locating prey, it holds the animal with the foot and drills a round hole in the shell. The moon snails are voracious; young animals eat more than a third of their weight in clams each week. Some worms also are predatory burrowers; so are a few starfish. But for most predators, continuous burrowing consumes more energy than would be supplied by the prey thus found. Most of the burrowers in sand are passive feeders, digging only enough to establish a temporary or permanent home in which to lie while straining food from the water or sucking up detritus that accumulates on the sea bottom.

The rising tide sets in action a system of living filters through which prodigious quantities of water are strained. Buried mollusks push up their siphons through the sand to draw the incoming water through their bodies. Worms lying in U-shaped parchment tubes begin to pump, drawing the water in through one end of the tube, expelling it through the other. The incoming stream brings food and oxygen; the outgoing has been depleted of much of the food and bears away the organic wastes of the worm. Small crabs spread the feathery nets of their antennae like cast-nets to bring in food.

With the tide, predators come from offshore. A blue crab dashes out of the surf to seize a fat mole crab that is in the act of spreading its antennae to filter the backwash of a receding wave. Clouds of salt-water minnows move in with the tide, searching for the small amphipods of the upper beach. Launce, or sand eels, dart through the shallow water seeking copepods or fish fry; sometimes the launce are pursued by the shadowy forms of larger fish.

As the tide falls much of this extraordinary activity slackens. There is less eating and being eaten. In the wet sands, however, some animals can continue to eat even after the tide has receded. Lugworms can continue their work of passing sand through their bodies for the sake of the scraps of nutriment they contain. Heart urchins and sand dollars, lying in saturated sand, continue to sort out bits of food. But over most of the sands there is a lull of repletion—of waiting for the turn of the tide.

Although there are many places where, on quieter shores and

protected shoals, such richness of life may be found, certain ones live most clearly in my memories. On one of the sea islands of Georgia is a great beach that is visited only by the most gentle surf, although it looks straight across to Africa. Storms usually pass it by, for it lies well inside the long, incurving arc of coast that swings between the Capes of Fear and Canaveral, and the prevailing winds are such that no heavy swells roll in upon it. The texture of the beach itself is unusually firm because of a mixture of mud and clay with the sand; permanent holes and burrows can be dug in it, and the streaming tidal currents carve little ripple marks that remain after the tide goes out, looking like a miniature model of the sea's waves. These sand ripples hold small food particles dropped by the currents, providing a store to be drawn on by detritus feeders. The slope of the beach is so gentle that, when the tide falls to its lowest ebb, a quarter of a mile of sand is exposed between the high-tide line and the low. But this broad sand flat is not a perfectly even plain, for winding gullies wander across it, like creeks across the land, holding a remnant of water from the last high tide and providing a living place for animals that cannot endure even a temporary withdrawal of the water.

It was in this place that I once found a large "bed" of sea pansies at the very edge of the tide. The day was heavily overcast, a fact that accounted for their being exposed. On sunny days I never saw them there, although undoubtedly they were just under the sand, protecting themselves from the drying rays of the sun.

But the day I saw them the pink and lavender flower faces were lifted so that they were exposed at the surface of the sand, though so slightly that one could easily pass them by unnoticed. Seeing them— even recognizing them for what they were—there was a sense of incongruity in finding what looked so definitely flowerlike here at the edge of the sea.

These flattened, heart-shaped sea pansies, raised on short stems above the sand, are not plants but animals. They belong to the same general group of simple beings as the jellyfish, sea anemones, and corals, but to find their nearest relatives one would have to desert the shore and go down to some deeplying offshore bottom where, as fernlike growths in a strange animal forest, the sea pens thrust long stalks into the soft ooze.

Each sea pansy growing here at the edge of the tide is the product of a minute larva that once dropped from the currents to this shore. But through the extraordinary course of its development it has ceased to be that single being of its origin and has become instead a group or colony of many individuals, bound together into a whole of flowerlike form. The various individuals or polyps all have the shape of little tubes embedded in the fleshy substance of the colony. But some of the tubes bear tentacles and look like very small sea anemones; these capture food for the colony, and in the proper season form reproductive cells. Other tubes lack tentacles; these are the engineers of the colony, attending to the functions of water-intake and control. A hydraulic system of changing water pressure controls the movements of the colony; as the stem is made turgid it may be thrust down into the sand, drawing the main body after it.

As the rising tide streams over the flattened shapes of the sea pansies, all the tentacles of the feeding polyps are thrust up, reaching for the living motes that dance in the water—the copepods, the diatoms, the fish larvae small and tenuous as threads.

And at night the shallow water, rippling gently over these flats, must glow softly with hundreds of little lights marking out the zone where the sea pansies live, in a serpentine line of gleaming points, just as lights seen from an airplane at night wander across the dark landscape and show the path of settlement along a highway. For the sea pansies, like their deep-sea relatives, are beautifully luminescent.

In season, the tide sweeping over these flats carries many small, pear-shaped, swimming larvae from which new colonies of pansies will develop. In past ages, the currents that traversed the open water then separating North and South America carried such larvae, which established themselves on the Pacific coast, north to Mexico and south to Chile. Then a bridge of land rose between the American continents, closing the water highway. Today the presence of sea pansies on both Atlantic and Pacific coasts is one of the living reminders of that past geologic time when North and South America were separated, and sea creatures passed freely from one ocean to the other.

In that liquescent sand at the edge of the low tide, I often saw small bubblings and boilings under the surface as one or another of the sand dwellers slipped in or out of its hidden world.

There were sand dollars, or keyhole urchins, thin as wafers. As one of them buried itself the forward edge slipped obliquely into the sand, passing with effortless ease from the world of sunlight and water into those dim regions of which my senses knew nothing. Internally, the shells are strengthened for burrowing, and against the force of surf, by supporting pillars that occupy most of the region between upper and lower shells except in the center of the disc. The surface of the animal is covered with minute spines, soft as felt. The spines shimmered in the sunlight as their waving movements set up currents that kept the sand grains in motion and eased the passage of the creature from water into earth. On the back of the disc was dimly marked out a design like a five-petaled flower. Repeating the meaning and the symbolism of the number five—the sign of the echinoderms— were five holes perforating the flat disc. As the animal progressed just under the shifting film of surface sand, grains moved up from the under side through the holes, aiding its forward movement and spreading a concealing veil of sand over its body.

The sand dollars shared their dark world with other echinoderms. Down in the wet sand lived heart urchins, which one never sees at the surface until the thin little boxes that once contained them are found by the tide and carried in to the beach, to be blown about by the wind and left at last in the litter of the high-tide line. The oddly shaped heart urchins lay in chambers six inches or more below the surface of the sand, keeping open for themselves channels lined with sticky mucus; through these they reached up to the floor of the shallow sea, finding diatoms and other particles of food among the sand grains.

And sometimes a starlike pattern twinkled in that firmament of sand, proclaiming that one of the sand-dwelling starfishes lay below, marking out its image by the flow of water currents, as the animal drew sea water through its body for respiration, expelling it through many pores on its upper surface. If the sand was disturbed, the astral image trembled and faded, like a star disappearing in mist, as the animal glided away rapidly, paddling through the sand with flattened tube feet.

Walking back across the flats of that Georgia beach, I was always aware that I was treading on the thin rooftops of an underground city. Of the inhabitants themselves little or nothing was visible. There were the chimneys and stacks and ventilating pipes of underground dwellings, and various passages and runways leading down into darkness. There were little heaps of refuse that had been brought up to the surface as though in an attempt at some sort of civic sanitation. But the inhabitants remained hidden, dwelling silently in their dark, incomprehensible world.

The most numerous inhabitants of this city of burrowers were the ghost shrimps. Their holes were everywhere over the tidal flat, in diameter considerably smaller than a lead pencil, and surrounded by a little pile of fecal pellets. The pellets accumulate in great quantity because of the shrimp's way of life; it must eat an enormous amount of sand and mud to obtain the food that is mixed with this indigestible material. The holes are the visible entrances to burrows that extend down several feet into the sand—long, nearly vertical passageways from which other tunnels lead off, some continuing down into the dark, damp basement of this shrimp city, others leading up to the surface as though to provide emergency exit doors.

The owners of the burrows did not show themselves unless I tricked them into it by dropping sand grains, a few at a time, into their entrance halls. The ghost shrimp is a curiously formed creature with a long slender body. It seldom goes abroad and so has no need of a hard protective skeleton; it is covered, instead, with a flexible cuticle suited to the narrow tunnel in which it must be able to dig and turn about. On the under side of its body are several pairs of flattened appendages that beat continually to force a current of water through the burrow, for in the deep sand layers the oxygen supply is poor, and aerated water must be drawn down from above. When the tide comes in, the ghost shrimps go up to the mouths of their burrows and begin their work of sifting the sand grains for bacteria, diatoms, and perhaps larger particles of organic detritus. The food is brushed out of the sand by means of little hairs on several of the appendages, and is then transferred to the mouth.

Few of those who build permanent homes in this underground city of sand live by themselves. On the Atlantic coast, the ghost shrimp regularly gives lodging to a small rotund crab, related to the species often found in oysters. The pea crab, Pinnixa, finds in the well-aerated burrow of the shrimp both shelter and a steady supply of food. It strains food out of the water currents that flow through the burrow, using little feathery outgrowths of its body as nets. On the California coast the ghost shrimp shelters as many as ten different species of animals. One is a fish—a small goby—that uses the burrow as a casual refuge while the tide is out, roaming through the passageways of the shrimp's home and pushing past the owner when necessary. Another is a clam that lives outside the burrow but thrusts its siphons through the walls and takes food from the water circulating through the tunnel. The clam has short siphons and in ordinary circumstances would have to live just under the surface of the sand to reach water and its food supply; by establishing connection with the shrimp's burrow it is able to enjoy the protective advantages of living at a deeper level.

On the muddier parts of these same Georgia flats the lugworm lives, its presence marked by round black domes, like low volcanic cones. Wherever the lugworms occur, on shores of America and Europe, their prodigious toil leavens and renews the beaches and keeps the amount of decaying organic matter in proper balance. Where they are abundant, they may work over in a year nearly two thousand tons of soil per acre. Like its counterpart on land, the earthworm, the lugworm passes quantities of soil through its body. The food in decaying organic debris is absorbed by its digestive tract; the sand is expelled in neat, coiled castings that betray the presence of the worm. Near every dark cone, a small, funnel-shaped depression appears in the sand. The worm lies within the sand in the shape of the letter U, the tail under the cone, the head under the depression. When the tide rises, the head is thrust out to feed.

Other signs of the lugworm appear in midsummer—large, translucent, pink sacs, each bobbing about in the water like a child's balloon, with one end drawn down into the sand. These compact masses of jelly are the egg masses of the worm, within each of which as many as 300,000 young are undergoing development.

Vast plains of sand are continually worked over by these and other marine worms. One—the trumpet worm—uses the very sand that contains its food to make a cone-shaped tube for the protection of its soft body in tunneling. One may sometimes see the living trumpet worm at work, for it allows its tube to project slightly above the surface. It is much more common, however, to find the empty tubes in the tidal debris. Despite their fragile appearance, they remain intact long after their architects are dead—natural mosaics of sand, one grain thick, the building stones fitted together with meticulous care.

A Scot named A. T. Watson once spent many years studying the habits of this worm. Because tube-building goes on under ground, he found it almost impossibly difficult to observe the fitting into place and cementing of sand grains until he hit upon the idea of collecting very young larvae, which could live and be observed in a thin layer of sand in the bottom of a laboratory dish. The building of the tube was begun soon after the larvae had ceased to swim about and had settled on the bottom of the dish. First each secreted a membranous tube about itself. This was to become the inner lining of the cone, and the foundation for the sand-grain mosaic. These young larvae had only two tentacles, which they used to collect grains of sand and pass them to the mouth. There the grains were rolled about experimentally, and if found suitable, were deposited on the chosen spot at the edge of the tube. Then a little fluid was expelled from the cement gland, after which the worm rubbed certain shield-like structures over the tube as though to smooth it.

"Each tube," wrote Watson, "is the life work of the tenant, and is most beautifully built with grains of sand, each grain placed in position with all the skill and accuracy of a human builder ... The moment when an exact fit has been obtained is evidently ascertained by an exquisite sense of touch. On one occasion I saw the worm slightly alter (before cementing) the position of a sand grain which it had just deposited."

The tubes serve to house the owners during a lifetime of subterranean tunneling, for like the lugworm, this species finds its food in the subsurface sands. The digging organs, like the tubes, belie their fragile appearance. They are slender, sharp-pointed bristles arranged in two groups, or "combs," which look fantastically impractical. We could easily believe that someone, in whimsical mood, had cut them out of shining golden foil, fringing the margins with repeated snips of the scissors to fashion a Christmas tree ornament.

I have watched the worms at work, in a miniature world of sand and sea created for them in my laboratory. Even in a thin layer of sand in a glass bowl, the combs are used with a sturdy efficiency that reminds one of a bulldozer. The worm emerges slightly from the tube, thrusts the combs into the sand, scoops up a load and throws it over its shoulder, as it were; than it seems to scrape the shovel blades clean by drawing them back over the edge of the tube. The whole thing is done with vigor and dispatch, with motions alternately to right and left. The golden shovels loosen the sand and allow the soft, food-gathering tentacles to explore among the grains, and bring to the mouth the food they discover.

Down along the line of barrier islands that stands between the mainland and the sea, the waves have cut inlets through which the tides pour into the bays and sounds behind the islands. The seaward shores of the islands are bathed by coastwise currents carrying their loads of sand and silt, mile after mile. In the confusion of meeting the tides that are racing to or from the inlets, the currents slacken and relax their hold on some of the sediments. So, off the mouths of many of the inlets, lines of shoals make out to sea—the wrecking sands of Diamond Shoal and Frying Pan Shoal and scores of others, named or nameless. But not all of the sediments are so deposited. Many are seized by the tides and swept through the inlets, only to be dropped in the quieter waters inside. Within the capes and the inlet mouths, in the bays and sounds, the shoals build up. Where they exist the searching larvae or young of sea creatures find them—creatures whose way of life requires quiet and shallow water.

Within the shelter of Cape Lookout there are such shoals reaching upward to the surface, emerging briefly into sun and air for the interval of the low tide, then sinking again into the sea. They are seldom crossed by heavy surf, and while the tidal currents that swirl over or around them may gradually alter their shape and extent—today borrowing some of their substance, tomorrow repaying it with sand or silt brought from other areas—they are on the whole a stable and peaceful world for the animals of the sands.

Some of the shoals bear the names of the creatures of air and water that visit them—Shark, Sheepshead, Bird. To visit Bird Shoal, one goes out by boat through channels winding through the Town Marsh of Beaufort and comes ashore on a rim of sand held firm by the deep roots of beach grasses—the landward border of the shoal. The burrows of thousands of fiddler crabs riddle the muddy beach on the side facing the marshes. The crabs shuffle across the flats at the approach of an intruder, and the sound of many small chitinous feet is like the crackling of paper. Crossing the ridge of sand, one looks out over the shoal. If the tide still has an hour or two to fall to its ebb, one sees only a sheet of water shimmering in the sun.

On the beach, as the tide falls, the border of wet sand gradually retreats toward the sea. Offshore, a dull velvet patch takes form on the shining silk of the water, like the back of an immense fish slowly rolling out of the sea, as a long streak of sand begins to rise into view.

On spring tides the peak of this great sprawling shoal rises farther out of the water and is exposed longer; on the neaps, when the tidal pulse is feeble and the water movements sluggish, the shoal remains almost hidden, with a thin sheet of water rippling across it even at the low point of the ebb. But on any low tide of the month, in calm weather, one is able to wade out from the sand-dune rim over immense areas of the shoal, in water so shallow and so glassy clear that every detail of the bottom lies revealed.

Even on moderate tides I have gone so far out that the dry sand rim seemed far away. Then deep channels began to cut across the outlying parts of the shoal. Approaching them, I could see the bottom sloping down out of crystal clarity into a green that was dull and opaque. The steepness of the slope was accentuated when a little school of minnows flickered across the shallows and down into the darkness in a cascade of silver sparks. Larger fish wandered in from the sea along these narrow passages between the shoals. I knew there were beds of sun ray clams down there on the deeper bottoms, with whelks moving down to prey on them. Crabs swam about or buried themselves to the eyes in the sandy bottoms; then behind each crab two small vortices appeared in the sand, marking the respiratory currents drawn in through the gills.

Where water—even the shallowest of layers—covered the shoal, life came out of hiding. A young horseshoe crab hurried out into deeper water; a small toadfish huddled down in a clump of eelgrass and croaked an audible protest at the foot of a strange visitor in his world, where human beings seldom intrude. A snail with neat black spirals around its shell and a matching black foot and black, tubular siphons—a banded tulip shell—glided rapidly over the bottom, tracing a clear track across the sand.

Here and there the sea grasses had taken hold—those pioneers among the flowering plants that are venturing out into salt water. Their flat leaf blades pushed up through the sand and their interlacing roots lent firmness and stability to the bottom. In such glades I found colonies of a curious, sand-dwelling sea anemone. Because of their structure and habits, anemones require some firm support to grip while reaching into the water for food. In the north (or wherever there is firm bottom) they grasp the rocks; here they gain the same end by pushing down into the sand until only the crown of tentacles remains above the surface. The sand anemone burrows by contracting the downward-pointing end of its tube and thrusting downward, then as a slow wave of expansion travels up the body, the creature sinks into the sand. It was strange to see the soft tentacle-clusters of the anemones flowering here in the midst of the sands, for anemones seem always to belong to the rocks; yet buried in this firm bottom doubtless they were as secure as the great plumose anemone blooming on the wall of a Maine tide pool.

Here and there over the grassy parts of the shoal the twin chimneys of the parchment worm's tubes protruded slightly above the sand. The worm itself lives always underground, in a U-shaped tube whose narrowed tips are the animal's means of contact with the sea. Lying in its tube, it uses fanlike projections of the body to keep a current of water streaming through the dark tunnel of its home, bringing it the minute plant cells that are its principal food, carrying away its waste products and in season the seeds of a new generation.

The whole life of the worm is so spent except for the short period of larval life at sea. The larva soon ceases to swim and, becoming sluggish, settles to the bottom. It begins to creep about, perhaps finding food in the diatoms lying in the troughs of the sand ripples. As it creeps it leaves a trail of mucus. After perhaps a few days the young worm begins to make short, mucus-coated tunnels, burrowing into thick clumps of diatoms mixed with sand. From such a simple tunnel, extending perhaps several times the length of its body, the larva pushes up extensions to the surface of the sand, to create the U-shape. All later tunnels are the result of repeated remodelings and extensions of this one, to accommodate the growing body of the worm. After the worm dies the limp, empty tubes are washed out of the sand and are common in the flotsam of the beach.

At some time almost all parchment worms acquire lodgers— the small pea crabs whose relatives inhabit the burrows of the ghost shrimps. Often the association is for life. The crabs, lured by the continuous stream of food-laden water, enter the worm tube while young, but soon become too large to leave by the narrow exits. Nor does the worm itself actually leave its tube, although occasionally one sees a specimen with a regenerated head or tail—mute evidence that it may emerge enough to tempt a passing fish or crab. Against such attacks it has no defense, unless the weird blue-white light that illuminates its whole body when disturbed may sometimes alarm an enemy.

Other little protruding chimneys raised above the surface of the shoal belonged to the plumed or decorator worm, Diopatra. These occurred singly, instead of in pairs. They were curiously adorned with bits of shell or seaweed that effectively deceived the human eye, and were but the exposed ends of tubes that sometimes extended down into the sand as much as three feet. Perhaps the camouflage is effective also against natural enemies, yet to collect the materials that it glues to all exposed parts of its tube, the worm has to expose several inches of its body. Like the parchment worm, it is able to regenerate lost tissues as a defense against hungry fish.

As the tide ebbed away, the great whelks could be seen here and there gliding about in search of their prey, the clams that lay buried in the sands, drawing through their bodies a stream of sea water and filtering from it microscopic plants. Yet the search of the whelks was not an aimless one, for their keen taste sense guided them to invisible streams of water pouring from the outlet siphons of the clams. Such a taste trail might lead to a stout razor clam, whose shells afford only the scantiest covering for its bulging flesh, or to a hard-shell clam, with tightly closed valves. Even these can be opened by a whelk, which grips the clam in its large foot and, by muscular contractions, delivers a series of hammer blows with its own massive shell.

Nor does the cycle of life—the intricate dependence of one species upon another—end there. Down in dark little dens of the sea floor live the enemies of the whelks, the stone crabs of massive purplish bodies and brightly colored crushing claws that are able to break away the whelk's shell, piece by piece. The crabs lurk in caves among the stones of jetties, in holes eroded out of shell rock, or in man-made homes such as old, discarded automobile tires. About their lairs, as about the abodes of legendary giants, lie the broken remains of their prey.

If the whelks escape this enemy, another comes by air. The gulls visit the shoal in numbers. They have no great claws to crush the shells of their victims, but some inherited wisdom has taught them another device. Finding an exposed whelk, a gull seizes it and carries it aloft. It seeks a paved road, a pier, or even the beach itself, soars high into the air and drops its prey, instantly following it earthward to recover the treasure from among the shattered bits of shell.

Coming back over the shoal, I saw spiraling up out of the sand, over the edge of a green undersea ravine, a looped and twisted strand—a tough string of parchment on which were threaded many scores of little purse-shaped capsules. This was the egg string of a female whelk, for it was June, and the spawning time of the species. In all the capsules, I knew, the mysterious forces of creation were at work, making ready thousands of baby whelks, of which perhaps hundreds would survive to emerge from the thin round door in the wall of each capsule, each a tiny being in a miniature shell like that of its parents.

Where the waves roll in from the open Atlantic, with no outlying islands or curving arm of land to break the force of their attack on the beach, the area between the tide lines is a difficult one for living things. It is a world of force and change and constant motion, where even the sand acquires some of the fluidity of water. These exposed beaches have few inhabitants, for only the most specialized creatures can live on sand amid heavy surf.

Animals of open beaches are typically small, always swift-moving. Theirs is a strange way of life. Each wave breaking on the beach is at once their friend and enemy; though it brings food, it threatens to carry them out to sea in its swirling backwash. Only by becoming amazingly proficient in rapid and constant digging can any animal exploit the turbulent surf and shifting sand for the plentiful food supplies brought in by the waves.

One of the successful exploiters is the mole crab, a surf-fisher who uses nets so efficient that they catch even microorganisms adrift in the water. Whole cities of mole crabs live where the waves are breaking, following the flood tide shoreward, retreating toward the sea on the ebb. Several times during the rising of a tide, a whole bed of them will shift its position, digging in again farther up the beach in what is probably a more favorable depth for feeding. In this spectacular mass movement, the sand area suddenly seems to bubble, for in a strangely concerted action, like the flocking of birds or the schooling of fish, the crabs all emerge from the sand as a wave sweeps over them. In the rush of turbulent water they are carried up the beach; then, as the wave's force slackens, they dig into the sand with magical ease, by means of a whirling motion of the tail appendages. With the ebbing of the tide, the crabs return toward the low-water mark, again making the journey in several stages. If by mischance a few linger until the tide has dropped below them, these crabs dig down several inches into the wet sand and wait for the return of the water.

As the name suggests, there is something mole-like in these small crustaceans, with their flattened, pawlike appendages. Their eyes are small and practically useless. Like all others who live within the sands the crabs depend less on sight than on the sense of touch, made wonderfully effective by the presence of many sensory bristles. But without the long, curling, feathery antennae, so efficiently constructed that even small bacteria become entangled in their strands, the mole crab could not survive as a fisher of the surf. In preparing to feed, the crab backs down into the wet sand until only the mouth parts and the antennae are exposed. Although it lies facing the ocean, it makes no attempt to take food from the incoming surf. Rather, it waits until a wave has spent its force on the beach and the backwash is draining seaward. When the spent wave has thinned to a depth of an inch or two, the mole crab extends its antennae into the streaming current. After "fishing" for a moment, it draws the antennae through the appendages surrounding its mouth, picking off the captured food. And again in this activity there is a curious display of group behavior, for when one crab thrusts up its antennae, all the others of the colony promptly follow its example.

It is an extraordinary thing to watch the sand come to life if one happens to be wading where there is a large colony of the crabs. One moment it may seem uninhabited. Then, in that fleeting instant when the water of a receding wave flows seaward like a thin stream of liquid glass, there are suddenly hundreds of little gnome-like faces peering through the sandy floor—beady-eyed, long-whiskered faces set in bodies so nearly the color of their background that they can barely be seen. And when, almost instantly, the faces fade back into invisibility, as though a host of strange little troglodytes had momentarily looked out through the curtains of their hidden world and as abruptly retired within it, the illusion is strong that one has seen nothing except in imagination—that there was merely an apparition induced by the magical quality of this world of shifting sand and foaming water.

Since their food-gathering activities keep them in the edge of the surf, mole crabs are exposed to enemies from both land and water—birds that probe in the wet sand, fish that swim in with the tide, feeding in the rising water, blue crabs darting out of the surf to seize them. So the mole crabs function in the sea's economy as an important link between the microscopic food of the waters and the large, carnivorous predators.

Even though the individual mole crab may escape the larger creatures that hunt the tide lines, the span of life is short, comprising a summer, a winter, and a summer. The crab begins life as a minute larva hatched from an orange-colored egg that has been carried for months by the mother crab, one of a mass firmly attached beneath her body. As the time for hatching nears, the mother foregoes the feeding movements up and down the beach with the other crabs and remains near the zone of the low tide, so avoiding the danger of stranding her offspring on the sands of the upper beach.

When it escapes from the protective capsule of the egg, the young larva is transparent, large-headed, and large-eyed as are all crustacean young, weirdly adorned with spines. It is a creature of the plankton, knowing nothing of life in the sands. As it grows it molts, shedding the vestments of its larval life. So it reaches a stage in which, although still swimming in larval fashion with waving motions of its bristled legs, it now seeks the bottom in the turbulent surf zone, where the waves stir and loosen the sand. Toward the summer's end there is another molt, this time bringing transformation to the adult stage, with the feeding behavior of the adult crabs.

During the protracted period of larval life, many of the young mole crabs have made long coastwise journeys in the currents, so that their final coming ashore (if they have survived the voyage) may be far from the parental sands. On the Pacific coast, where strong surface currents flow seaward, Martin Johnson found that great numbers of the crab larvae are carried out over oceanic depths, doomed to certain destruction unless they chance to find their way into a return current. Because of the long larval life, some of the young crabs are carried as far as 200 miles offshore. Perhaps in the prevailing coastwise current of Atlantic shores they travel even farther.

With the coming of winter the mole crabs remain active. In the northern part of their range, where frost bites deep into the sands and ice may form on the beaches, they go out beyond the low-tide zone to pass the cold months where a fathom or more of insulating water lies between them and the wintry air. Spring is the mating season and by July most or all of the males hatched the preceding summer have died. The females carry their egg masses for several months until the young hatch; before winter all of these females have died and only a single generation of the species remains on the beach.

The only other creatures regularly at home between the tide lines of wave-swept Atlantic beaches are the tiny coquina clams. The life of the coquinas is one of extraordinary and almost ceaseless activity. When washed out by the waves, they must dig in again, using the stout, pointed foot as a spade to thrust down for a firm grip, after which the smooth shell is pulled rapidly into the sand. Once firmly entrenched, the clam pushes up its siphons. The intake siphon is about as long as the shell and flares widely at the mouth. Diatoms and other food materials brought in or stirred from the bottom by waves are drawn down into the siphon.

Like the mole crabs, the coquinas shift higher or lower on the beach in mass movements of scores or hundreds of individuals, perhaps to take advantage of the most favorable depth of water. Then the sand flashes with the brightly colored shells as the clams emerge from their holes and let the waves carry them. Sometimes other small burrowers move with the coquinas among the waves—companies of the little screw shell, Terebra, a carnivorous snail that preys on the coquina. Other enemies are sea birds. The ring-billed gulls hunt the clams persistently, treading them out of the sand in shallow water.

On any particular beach, the coquinas are transient inhabitants; they seem to work an area for the food it provides, and then move on. The presence on a beach of thousands of the beautifully variegated shells, shaped like butterflies and crossed by radiating bands of color, may mark only the site of a former colony.

Being only briefly and sporadically possessed by the sea in those recurrent periods of the tides' farthest advance, the high-tide zone on any shore has in its own nature something of the land as well as of the sea. This intermediate, transitional quality pervades not only the physical world of the upper beach but also its life. Perhaps the ebb and flow of the tides has accustomed some of the intertidal animals, little by little, to living out of water; perhaps this is the reason there are among the inhabitants of this zone some who, at this moment of their history, belong neither to the land nor entirely to the sea.

The ghost crab, pale as the dry sand of the upper beaches it inhabits, seems almost a land animal. Often its deep holes are back where the dunes begin to rise from the beach. Yet it is not an air-breather; it carries with it a bit of the sea in the branchial chamber surrounding its gills, and at intervals must visit the sea to replenish the water. And there is another, almost symbolic return. Each of these crabs began its individual life as a tiny creature of the plankton; after maturity and in the spawning season, each female enters the sea again to liberate her young.

If it were not for these necessities, the lives of the adult crabs would be almost those of true land animals. But at intervals during each day they must go down to the water line to wet their gills, accomplishing their purpose with the least possible contact with the sea. Instead of wading directly into the water, they take up a position a little above the place where, at the moment, most of the waves are breaking on the beach. They stand sideways to the water, gripping the sand with the legs on the landward side. Human bathers know that in any surf an occasional wave will tower higher than the others and run farther up the beach. The crabs wait, as if they also know this, and after such a wave has washed over them, they return to the upper beach.

They are not always wary of contact with the sea. I have a mental picture of one sitting astride a sea-oats stem on a Virginia beach, one stormy October day, busily putting into its mouth food particles that it seemed to be picking off the stem. It munched away, intent on its pleasant occupation, ignoring the great, roaring ocean at its back. Suddenly the foam and froth of a breaking wave rolled over it, hurling the crab from the stem and sending both slithering up the wet beach. And almost any ghost crab, hard pressed by a person trying to catch it, will dash into the surf as though choosing a lesser evil. At such times they do not swim, but walk along on the bottom until their alarm has subsided and they venture out again.

Although on cloudy days and even occasionally in full sunshine the crabs may be abroad in small numbers, they are predominantly hunters of the night beaches. Drawing from the cloak of darkness a courage they lack by day, they swarm boldly over the sand. Sometimes they dig little temporary pits close to the water line, in which they lie watching for what the sea may bring them.

The individual crab in its brief life epitomizes the protracted racial drama, the evolutionary coming-to-land of a sea creature. The larva, like that of the mole crab, is oceanic, becoming a creature of the plankton once it has hatched from the egg that has been incubated and aerated by the mother. As the infant crab drifts in the currents it sheds its cuticle several times to accommodate the increasing size of its body; at each molt it undergoes slight changes of form. Finally the last larval stage, called the megalops, is reached. This is the form in which all the destiny of the race is symbolized, for it—a tiny creature alone in the sea—must obey whatever instinct drives it shoreward, and must make a successful landing on the beach. The long processes of evolution have fitted it to cope with its fate. Its structure is extraordinary when compared with like stages of closely related crabs. Jocelyn Crane, studying these larvae in various species of ghost crabs, found that the cuticle is always thick and heavy, the body rounded. The appendages are grooved and sculptured so that they may be folded down tightly against the body, each fitting precisely against the adjacent ones. In the hazardous act of coming ashore, these structural adaptations protect the young crab against the battering of the surf and the scraping of sand.

Once on the beach, the larva digs a small hole, perhaps as protection from the waves, perhaps as a shelter in which to undergo the molt that will transform it into the shape of the adult. From then on, the life of the young crab is a gradual moving up the beach. When small it digs its burrows in wet sand that will be covered by the rising tide. When perhaps half grown, it digs above the high-tide line; when fully adult it goes well back into the upper beach or even among the dunes, attaining then the farthest point of the landward movement of the race.

On any beach inhabited by ghost crabs, their burrows appear and disappear in a daily and seasonal rhythm related to the habits of the owners. During the night the mouths of the burrows stand open while the crabs are out foraging on the beach. About dawn the crabs return. Whether each goes, as a rule, to the burrow it formerly occupied or merely to any convenient one is uncertain—the habit may vary with locality, the age of the crab, and other changing conditions.

Most of the tunnels are simple shafts running down into the sand at an angle of about forty-five degrees, ending in an enlarged den. Some few have an accessory shaft leading up from the chamber to the surface. This provides an emergency exit to be used if an enemy— perhaps a larger and hostile crab-comes down the main shaft. This second shaft usually runs to the surface almost vertically. It is farther away from the water than the main tunnel, and may or may not break through the surface of the sand.

The early morning hours are spent repairing, enlarging, or improving the burrow selected for the day. A crab hauling up sand from its tunnel always emerges sideways, its load of sand carried like a package under the legs of the functional rear end of the body. Sometimes, immediately on reaching the burrow mouth, it will hurl the sand violently away and flash back into the hole; sometimes it will carry it a little distance away before depositing it. Often the crabs stock their burrows with food and then retire into them; nearly all crabs close the tunnel entrances about midday.

All through the summer the occurrence of holes on the beach follows this diurnal pattern. By autumn most of the crabs have moved up to the dry beach beyond the tide; their holes reach deeper into the sand as though their owners were feeling the chill of October. Then, apparently, the doors of sand are pulled shut, not to be opened again until spring. For the winter beaches show no sign either of the crabs or of their holes—from dime-sized youngsters to full-grown adults, all have disappeared, presumably into the long sleep of hibernation. But, walking the beach on a sunny day in April, one will see here and there an open burrow. And presently a ghost crab in an obviously new and shiny spring coat may appear at its door and very tentatively lean on its elbows in the spring sunshine. If there is a lingering chill in the air, it will soon retire and close its door. But the season has turned, and under all this expanse of upper beach, crabs are awakening from their sleep.

Like the ghost crab, the small amphipod known as the sand hopper or beach flea portrays one of those dramatic moments of evolution, in which a creature abandons an old way of life for a new. Its ancestors were completely marine; its remote descendants, if we read its future aright, will be terrestrial. Now it is midway in the transition from a sea life to a land life.

As in all such transitional existences, there are strange little contradictions and ironies in its way of life. The sand hopper has progressed as far as the upper beach; its predicament is that it is bound to the sea, yet menaced by the very element that gave it life. Apparently it never enters the water voluntarily. It is a poor swimmer and may drown if long submerged. Yet it requires dampness and probably needs the salt in the beach sand, and so it remains in bondage to the water world.

The movements of the sand hoppers follow the rhythm of the tides and the alternation of day and night. On the low tides that fall during the dark hours, they roam far into the intertidal zone in search of food. They gnaw at bits of sea lettuce or eelgrass or kelp, their small bodies swaying with the vigor of their chewing. In the litter of the tide lines they find morsels of dead fish or crab shells containing remnants of flesh; so the beach is cleaned and the phosphates, nitrates, and other mineral substances are recovered from the dead for use by the living.

If low water has fallen late in the night, the amphipods continue their foraging until shortly before daybreak. Before light has tinged the sky, however, all of the hoppers begin to move up the beach toward the high-water line. There each begins to dig the burrow into which it will retreat from daylight and rising water. As it works rapidly, it passes back the grains of sand from one pair of feet to the next until, with the third pair of thoracic legs, it piles the sand behind it. Now and then the small digger straightens out its body with a snap, so that the accumulated sand is thrown out of the hole. It works furiously at one wall of the tunnel, bracing itself with the fourth and fifth pairs of legs, then turns and begins work on the opposite wall. The creature is small and its legs are seemingly fragile, yet the tunnel may be completed within perhaps ten minutes, and a chamber hollowed out at the end of the shaft. At its maximum depth this shaft represents as prodigious a labor as though a man, working with no tools but his hands, had dug for himself a tunnel about 60 feet deep.

The work of excavation done, the sand flea often returns to the mouth of its burrow to test the security of the entrance door, formed by the accumulation of sand from the deeper parts of the shaft. It may thrust out its long antennae from the mouth of the burrow, feeling the sand, tugging at the grains to draw more of them into the hole. Then it curls up within the dark snug chamber.

As the tide rises overhead, the vibrations of breaking waves and shoreward-pressing tides may come down to the little creature in its burrow, bringing a warning that it must stay within to avoid water and the dangers brought by water. It is less easy to understand what arouses the protective instinct to avoid daylight, with all the dangers of foraging shore birds. There can be little difference between day and night in that deep burrow. Yet in some mysterious way the beach flea is held within the safety of the sandy chamber until the two essential conditions again prevail on the beach—darkness and a falling tide.

Then it awakens from sleep, creeps up the long shaft, and pushes away the sand door. Once again the dark beach stretches before it, and a retreating line of white froth at the edge of the tide marks the boundary of its hunting grounds.

Each den that is dug with so much labor is merely a shelter for one night, or one tidal interval. After the low-tide feeding period, each hopper will dig itself a new refuge. The holes that we see on the upper beach lead down to empty burrows from which the former occupants have gone. An occupied burrow has its "door" closed, and so its location cannot easily be detected. On the sandy edge of the sea there is, then, the abundant life of protected beaches and shoals, the sparse life of surf-swept sands, and the pioneering life that has reached the high-tide line and seems poised in space and time for invasion of the land.

But the sands contain also the record of other lives. A thin net of flotsam is spread over the beaches—the driftage of ocean brought to rest on the shore. It is a fabric of strange composition, woven with tireless energy by wind and wave and tide. The supply of materials is endless. Caught in the strands of dried beach grass and seaweeds there are crab claws and bits of sponge, scarred and broken mollusk shells, old spars crusted with sea growths, the bones of fishes, the feathers of birds. The weavers use the materials at hand, and the design of the net changes from north to south. It reflects the kind of bottom offshore— whether rolling sand hills or rocky reefs; it subtly hints of the nearness of a warm, tropical current, or tells of the intrusion of cold water from the north. In the litter and debris of the beach there may be few living creatures, but there is the suggestion, the intimation of a million, million lives, lived in the sands nearby or brought to this place from far sea distances.

In the beach flotsam there are often strays from the surface waters of the open ocean, reminders of the fact that most sea creatures are the prisoners of the particular water masses they inhabit. When tongues of their native waters, driven by winds or drawn by varying temperature or salinity patterns, stray into unaccustomed territory, this drifting life is carried involuntarily with them.

In the several centuries that men of inquiring mind have been walking the world's shores many unknown sea animals have been discovered as strays from the open ocean in the flotsam of the tide lines. One such mysterious link between the open sea and the shore is the ramshorn shell, Spirula. For many years only the shell had been known—a small white spiral forming two or three loose coils. By holding such a shell to the light, one can see that it is divided into separate chambers, but seldom is there a trace of the animal that built and inhabited it. By 1912, about a dozen living specimens had been found, but still no one knew in what part of the sea the creature lived. Then Johannes Schmidt undertook his classic researches into the life history of the eel, crossing and recrossing the Atlantic and towing plankton nets at different levels from the surface down into depths perpetually black. Along with the glass-clear larvae of the eels that were the object of his search, he brought up other animals—among them many specimens of Spirula, which had been caught swimming at various depths down to a mile. In their zone of greatest abundance, which seems to lie between 900 and 1500 feet, they probably occur in dense schools. They are little squidlike animals with ten arms and a cylindrical body, bearing fins like propellers at one end. Placed in an aquarium, they are seen to swim with jerky, backward spurts of jet-propelled motion.

It may seem mysterious that the remains of such a deep-sea animal should come to rest in beach deposits, but the reason is, after all, not obscure. The shell is extremely light; when the animal dies and begins to decay, the gases of decomposition probably lift it toward the surface. There the fragile shell begins a slow drift in the currents, becoming a natural "drift bottle" whose eventual resting place is a clue not so much to the distribution of the species as to the course of the currents that bore it. The animals themselves live over deep oceans, perhaps most abundantly above the steep slopes that descend from the edges of the continents into the abyss. In such depths, they seem to occupy tropical and subtropical belts around the world. Now, in this little shell curved like the horn of a ram, we have one of the few persisting reminders of the days when great, spiral-shelled "cuttle fish" swarmed in the oceans of the Jurassic and earlier periods. All other cephalopods, except the pearly or chambered nautilus of the Pacific and Indian Oceans, have either abandoned their shells or converted them to internal remnants.

And sometimes, among the tidal debris, there appears a thin papery shell, bearing on its white surface a ribbed pattern like that which shore currents impress upon the sand. It is the shell of the paper nautilus or argonaut, an animal distantly related to an octopus, and like it having eight arms. The argonaut lives on the high seas, in both Atlantic and Pacific Oceans. The "shell" is actually an elaborate egg case or cradle secreted by the female for the protection of her young. It is a separate structure that she can enter or leave at will. The much smaller male (about one tenth the size of his mate) secretes no shell. He inseminates the female in the strange manner of some other cephalopods: one of his arms breaks off and enters the mantle cavity of the female, carrying a load of spermatophores. For a long while the male of this creature went unrecognized. Cuvier, a French zoologist of the early nineteenth century, was familiar with the detached arm but supposed it to be an independent animal, probably a parasitic worm. The argonaut is not the chambered or pearly nautilus of Holmes's famous poem. Although also a cephalopod, the pearly nautilus belongs to a different group and bears a true shell secreted by the mantle. It inhabits tropical seas, and like Spirula is a descendant of the great spiral-shelled mollusks that dominated the seas of Mesozoic times.

Storms bring in many strays from tropical waters. In a shell shop at Nags Head, North Carolina, I once attempted to buy the beautiful violet snail, Janthina. The proprietor of the shop refused to sell this, her only specimen. I understood why when she told me of finding the living Janthina on the beach after a hurricane, its marvelous float still intact, and the surrounding sand stained purple as the little animal tried, in its extremity, to use its only defense against disaster. Later I found an empty shell, light as thistledown, resting in a depression in the coral rock of Key Largo, where some gentle tide had laid it. I have never been so fortunate as my acquaintance at Nags Head, for I have never seen the living animal.

Janthina is a pelagic snail that drifts on the surface of the open ocean, hanging suspended from a raft of frothy bubbles. The raft is formed from mucus that the animal secretes; the mucus entraps bubbles of air, then hardens into a firm, clear substance like stiff cellophane. In the breeding season the snail fastens its egg capsules to the under side of the raft, which throughout the year serves to keep the little animal afloat.

Like most snails, Janthina is carnivorous; its prey is found among other plankton animals, including small jellyfishes, crustaceans, and even small goose barnacles. Now and then a swooping gull drops from the sky and takes a snail—but for the most part the bubble raft must be excellent camouflage, almost indistinguishable from a bit of drifting sea froth. There must be other enemies that come from below, for the blue-to-violet tints of the shell (which hangs below the raft) are the colors worn by many creatures that live at or near the surface film and need to conceal themselves from enemies looking up from below.

The strong northward flow of the Gulf Stream bears on its surface fleets of living sails—those strange coelenterates of the open sea, the siphonophores. Because of adverse winds and currents these small craft sometimes come into shallow water and are stranded on the beaches. This happens most often in the south, but the southern coast of New England also receives strays from the Gulf Stream, for the shallows west of Nantucket act as a trap to collect them. Among such strays, the beautiful azure sail of the Portuguese man-of-war, Physalia, is known to almost everyone, for so conspicuous an object can hardly be missed by any beach walker. The little purple sail, or by-the-wind sailor, Velella, is known to fewer, perhaps because of its much smaller size and the fact that once left on the beach it dries quickly to an object that is hard to identify. Both are typically inhabitants of tropical waters, but in the warmth of the Gulf Stream they may sometimes go all the way across to the coast of Great Britain, where in certain years they appear in numbers.

In life the oval float of Velella is a beautiful blue color, with a little elevated crest or sail passing diagonally across it. The disc is about an inch and a half long and half as wide. This is not one animal but a composite one, or colony of inseparably associated individuals— the multiple offspring of a single fertilized egg. The various individuals carry on separate functions. A feeding individual hangs suspended from the center of the float. Small reproductive individuals cluster around it. Around the periphery of the float, feeding individuals in the form of long tentacles hang down to capture the small fry of the sea.

A whole fleet of Portuguese men-of-war is sometimes seen from vessels crossing the Gulf Stream when some peculiarity of the wind and current pattern has brought together a number of them. Then one can sail for hours or days with always some of the siphonophores in sight. With, the float or sail set diagonally across its base, the creature sails before the wind; looking down into the clear water one can see the tentacles trailing far below the float. The Portuguese man-of-war is like a small fishing boat trailing a drift net, but its "net" is more nearly like a group of high-voltage wires, so deadly is the sting of the tentacles to almost any fish or other small animal unlucky enough to encounter them.

The true nature of the man-of-war is difficult to grasp, and indeed many aspects of its biology are unknown. But, as with Velella, the central fact is that what appears to be one animal is really a colony of many different individuals, although no one of them could exist independently. The float and its base are thought to be one individual; each of the long trailing tentacles another. The food-capturing tentacles, which in a large specimen may extend down for 40 or 50 feet, are thickly studded with nematocysts or stinging cells. Because of the toxin injected by these cells, Physalia is the most dangerous of all the coelenterates.

For the human bather, even glancing contact with one of the tentacles produces a fiery welt; anyone heavily stung is fortunate to survive. The exact nature of the poison is unknown. Some people believe there are three toxins involved, one producing paralysis of the nervous system, another affecting respiration, the third resulting in extreme prostration and death, if a large dose is received: In areas where Physalia is abundant, bathers have learned to respect it. On some parts of the Florida coast the Gulf Stream passes so close inshore that many of these coelenterates are borne in toward the beaches by onshore winds. The Coast Guard at Lauderdale-by-the-Sea and other such places, when posting reports of tides and water temperatures, often includes forecasts of the relative number of Physalias to be expected inshore.

Because of the highly toxic nature of the nematocyst poisons, it is extraordinary to find a creature that apparently is unharmed by them. This is the small fish Nomeus, which lives always in the shadow of a Physalia. It has never been found in any other situation. It darts in and out among the tentacles with seeming impunity, presumably finding among them a refuge from enemies. In return, it probably lures other fish within range of the man-of-war. But what of its own safety? Is it actually immune to the poisons? Or does it live an incredibly hazardous life? A Japanese investigator reported years ago that Nomeus actually nibbles away bits of the stinging tentacles, perhaps in this way subjecting itself to minute doses of the poison throughout its life and so acquiring immunity. But some recent workers contend that the fish has no immunity whatever, and that every live Nomeus is simply a very lucky fish.

The sail, or float, of a Portuguese man-of-war is filled with gas secreted by the so-called gas gland. The gas is largely nitrogen (85 to 91 per cent) with a small amount of oxygen and a trace of argon. Although some siphonophores can deflate the air sac and sink into deep water if the surface is rough, Physalia apparently cannot. However, it does have some control over the position and degree of expansion of the sac. I once had a graphic demonstration of this when I found a medium-size man-of-war stranded on a South Carolina beach. After keeping it overnight in a bucket of salt water, I attempted to return it to the sea. The tide was ebbing; I waded out into the chilly March water, keeping the Physalia in its bucket out of respect for its stinging abilities, then hurled it as far into the sea as I could. Over and over, the incoming waves caught it and returned it to the shallows. Sometimes with my help, sometimes without, it would manage to take off again, visibly adjusting the shape and position of the sail as it scudded along before the wind, which was blowing out of the south, straight up the beach. Sometimes it could successfully ride over an incoming wave; sometimes it would be caught and hustled and bumped along through thinning waters. But whether in difficulty or enjoying momentary success, there was nothing passive in the attitude of the creature. There was, instead, a strong illusion of sentience. This was no helpless bit of flotsam, but a living creature exerting every means at its disposal to control its fate. When I last saw it, a small blue sail far up the beach, it was pointed out to sea, waiting for the moment it could take off again.

Although some of the derelicts of the beach reflect the pattern of the surface waters, others reveal with equal clarity the nature of the sea bottom offshore. For thousands of miles from southern New England to the tip of Florida the continent has a continuous rim of sand, extending in width from the dry sand hills above the beaches far out across the drowned lands of the continental shelf. Yet here and there within this world of sand there are hidden rocky areas. One of these is a scattered and broken chain of reefs and ledges, submerged beneath the green waters off the Carolinas, sometimes close inshore, sometimes far out on the western edge of the Gulf Stream. Fishermen call them "black rocks" because the blackfish congregate around them.

The charts refer to "coral" although the closest reef-building corals are hundreds of miles away, in southern Florida.

In the 1940's, biologist divers from Duke University explored some of these reefs and found that they are not coral, but an outcropping of a soft claylike rock known as marl. It was formed during the Miocene many thousands of years ago, then buried under layers of sediment and drowned by a rising sea. As the divers described them, these submerged reefs are low-lying masses of rock sometimes rising a few feet above the sand, sometimes eroded away to level platforms from which swaying forests of brown sargassum grow. In deep fissures other algae find places of attachment. Much of the rock is smothered under curious sea growths, plant and animal. The stony coralline algae, whose relatives paint the low-tide rocks of New England a deep, old-rose hue, encrust the higher parts of the open reef and fill its interstices. Much of the reef is covered by a thick veneer of twisting, winding, limy tubes—the work of living snails and of tube-building worms, forming a calcareous layer over the old, fossil rock. Through the years the accumulation of algae and the growth of snail and worm tubes have added, little by little, to the structure of the reef.

Where the reef rock is free from crusts of algae and worm tubes, boring mollusks—date mussels, piddocks, and small boring clams—have drilled into it, scraping out holes in which they lodge, while feeding on the minute life of the water. Because of the firm support provided by the reef, gardens of color bloom in the midst of the drabness of shifting sand and silt. Sponges, orange or red or ocher, extend their branches into the currents that drift across the reef. Fragile, delicately branching hydroids rise from the rocks and from their pale "flowers," in season, tiny jellyfish swim away. Gorgonians are like tall wiry grasses, orange and yellow. And a curious shrubby form of moss animal or bryozoan lives here, the tough and gelatinous structure of its branches containing thousands of tiny polyps, which thrust out tentacled heads to feed. Often this bryozoan grows around a gorgonian, then appearing like gray insulation around a dark, wiry core.

Were it not for the reefs, none of these forms could exist on this sandy coast. But because, through the changing circumstances of geologic history, the old Miocene rocks are now cropping out on this shallow sea floor, there are places where the planktonic larvae of such animals, drifting in the currents, may end their eternal quest for solidity.

After almost any storm, at such places as South Carolina's Myrtle Beach, the creatures from the reefs begin to appear on the intertidal sands. Their presence is the visible result of a deep turbulence in the offshore waters, with waves reaching down to sweep violently over those old rocks that have not known the crash of surf since the sea drowned them, thousands of years ago. The storm waves dislodge many of the fixed and sessile animals and sweep off some of the free-living forms, carrying them away into an alien world of sandy bottoms, of waters shallowing ever more and more until there is no more water beneath them, only the sands of the beach.

I have walked these beaches in the biting wind that lingers after a northeast storm, with the waves jagged on the horizon and the ocean a cold leaden hue, and have been stirred by the sight of masses of the bright orange tree sponge lying on the beach, by smaller pieces of other sponges, green and red and yellow, by glistening chunks of "sea pork" of translucent orange or red or grayish white, by sea squirts like knobby old potatoes, and by living pearl oysters still gripping the thin branches of gorgonians. Sometimes there have been living starfish— the dark red southern form of the rock-dwelling Asterias. Once there was an octopus in distress on the wet sands where the waves had thrown it. But life was still in it; when I helped it out beyond the breakers it darted away.

Pieces of the ancient reef itself are commonly found on the sand at Myrtle Beach and presumably at any place where such reefs lie offshore. The marl is a dull gray cement-like rock, full of the borings of mollusks and sometimes retaining their shells. The total number of borers is always so great that one thinks how intense must be the competition, down on that undersea rock platform, for every available inch of solid surface, and how many larvae must fail to find a footing.

Another kind of "rock" occurs on the beach in chunks of varied size and perhaps even more abundantly than the marl. It has almost the structure of honeycomb taffy, being completely riddled with little twisting passageways. The first time one sees this on the beach, especially if it is half buried in sand, one might almost take it for one of the sponges, until investigation proves it to be hard as rock. It is not of mineral origin, however—it is built by small sea worms, dark of body and tentacled of head. These worms, living in aggregations of many individuals, secrete about themselves a calcareous matrix, which hardens to the firmness of rock. Presumably it thickly encrusts the reefs or builds up solid masses from a rocky floor. This particular kind of "worm rock" had not been known from the Atlantic coast until Dr. Olga Hartman identified my specimens from Myrtle Beach as "a matrix-building species of Dodecaceria" whose closest relatives are Pacific and Indian Ocean inhabitants. How and when did this particular species reach the Atlantic? How extensive is its range there? These and many other questions remain to be answered; they are one small illustration of the fact that our knowledge is encompassed within restricted boundaries, whose windows look out upon the limitless spaces of the unknown.

On the upper beach, beyond the zone where the flood tide returns the sea water twice daily, the sands dry out. Then they are subjected to excesses of heat; their arid depths are barren, with little to attract life, or even to make life possible. The grains of dry sands rub one against another. The winds seize them and drive them in a thin mist above the beach, and the cutting edge of this wind-driven sand scours the driftwood to a silver sheen, polishes the trunks of old derelict trees, and scourges the birds that nest on the beach.

But if this area has little life within itself, it is full of the reminders of other lives. For here above the high-tide line, all the empty shells of the mollusks come to rest. Visiting the beach that borders Shackleford Shoals in North Carolina or Florida's Sanibel Island, one could almost believe that mollusks are the only inhabitants of the sea's edge, for their enduring remains dominate the beach debris long after the more fragile remnants of crabs and sea urchins and starfish have been returned to the elements. First the shells were dropped low on the beach by the waves; then, tide by tide, they were moved up across the sands to the line of the highest of the high tides. Here they will remain, till buried in drifting sand or carried away in a wild carnival of storm surf.

From north to south the composition of the shell windrows changes, reflecting the changing communities of the mollusks. Every little pocket of gravelly sand that accumulates in favorable spots amid the rocks of northern New England is strewn with mussels and periwinkles. And when I think of the sheltered beaches of Cape Cod I see in memory the windrows of jingle shells being shifted gently by the tide, their thin, scale-like valves (how can they house a living creature?) gleaming with a satin sheen. The arched upper valve occurs more often in beach flotsam than the flat lower one, which is perforated by a hole for the passage of the strong byssus cord that attaches the jingle to a rock or to another shell. Silver, gold, and apricot are the colors of the jingles, set against the deep blue of the mussels that dominate these northern shores. And scattered here and there are the ribbed fans of the scallops and the little white sloops of the boat shells stranded on the beach. The boat shell is a snail with a curiously modified shell, having a little "half deck" on the lower surface. It often becomes attached to its fellows in chains of half a dozen or more individuals. Each boat shell is in its lifetime first male then female. In the chains of attached shells those at the bottom of the chain are always females, the upper animals males.

On the Jersey beaches and the coastal islands of Maryland and Virginia the massive structure of the shells and the lack of ornamental spines have a meaning—that the offshore world of shifting sand is deeply stirred by the endless processions of the waves that roll in on this coast. The thick shell of the surf clam is its defense against the force of the waves. These shores are strewn, too, with the heavy armaments of the whelks, and with the smooth globes of the moon snails.

From the Carolinas south the beach world seems to belong to the several species of arks, whose shells outnumber all others. Though variously shaped, their shells are stout, with long straight hinges. The ponderous ark wears a black, beardlike growth, or periostracum, heavy in fresh specimens, scanty or absent in beach-worn shells. The turkey wing is a gaily colored ark, with reddish bands streaking its yellowish shell. It, too, wears a thick periostracum, and lives down in deep offshore crevices, where it attaches itself to rocks or any other support by a strong line or byssus. While a few kinds of arks extend the range of these mollusks throughout New England (for example, the small transverse ark and the so-called bloody clam—one of the few mollusks that has red blood) it is on southern beaches that the group becomes dominant. On famed Sanibel Island on the west coast of Florida, where the variety of shells is probably greater than anywhere else on our Atlantic coast, the arks nevertheless make up about 95 per cent of the beach deposits.

The pen shells begin to appear in numbers on the beaches below Capes Hatteras and Lookout, but perhaps they, too, live in the most prodigious numbers on the Gulf coast of Florida. I have seen truckloads of them on the beach at Sanibel even in calm winter weather. In a violent tropical hurricane the destruction of this light-shelled mollusk is almost incredible. Sanibel Island presents about fifteen miles of beach to the Gulf of Mexico. On this strand, it has been estimated, about a million pen shells have been hurled by a single storm, having been torn loose by waves reaching down to bottoms lying as deep as 30 feet. The fragile shells of the pens are ground together in the buffeting of storm surf; many are broken, but even those not so destroyed have no way of returning to the sea, and so are doomed. As if knowing this, the commensal pea crabs that inhabit them creep out of the shells like the proverbial rats abandoning a sinking ship; they may be seen by the thousand swimming about in apparent bewilderment in the surf.

The pen shells spin anchoring byssus threads of golden sheen and remarkable texture; the ancients spun their cloth of gold from the byssus of the Mediterranean pens, producing a fabric so line and soft it could be drawn through a finger ring. The industry persists at Italian Taranto, on the Ionian Sea, where gloves and other small garments are woven of this natural fabric as curios or tourists' souvenirs.

The survival of an undamaged angel wing in the debris of the upper beach seems extraordinary, so delicately fragile does it appear. Yet these valves of purest white, when worn by the living animals, are capable of penetrating peat or firm clay. The angel wing is one of the most powerful of the boring clams and, having very long siphons with which to maintain communication with the sea water, is able to burrow deeply. I have dug for them in peat beds in Buzzards Bay, and have found them on beach exposures of peat on the coast of New Jersey, but their occurrence north of Virginia is local and rare.

This purity of color, this delicacy of structure are buried throughout life in a bank of clay, for the angel wing's beauty seems destined to be hidden from view until, after the death of the animal, the shells are released by the waves and carried to the beach. In its dark prison the angel wing conceals an even more mysterious beauty. Secure from enemies, hidden from all other creatures, the animal itself glows with a strange green light. Why? For whose eyes? For what reason?

Besides the shells, there are other objects in the beach flotsam that are mysterious in shape and texture. Flat, horny or shell-like discs of various shapes and sizes are the opercula of sea snails— the protective doors that close over the opening when the animal has withdrawn into its shell. Some opercula are round, some leaf-shaped, some like slender, curving daggers. (The "cat's eye" of the

South Pacific is the operculum of a snail, rounded on one surface and polished like a boy's marble.) The opercula of the various species are so characteristic in shape, material, and structure that they are a useful means of identifying otherwise difficult species.

The tidal flotsam abounds, too, in many little empty egg cases in which various sea creatures passed their first days of life. These are of various shapes and materials. The black "mermaid's purses" belong to one of the skates. They are flat, horny rectangles, with two long, curling prongs or tendrils extending from each end. With these the parent skate attaches the packet containing a fertilized egg to seaweeds on some offshore bottom. After the young skate matures and hatches, its discarded cradle is often washed up on the beach. Egg cases of the banded tulip shell remind one of the dried seed pods of a flower, a cluster of thin, parchment-like containers borne on a central stalk. Those of the channeled or the knobbed whelks are long, spiraling strings of little capsules, again parchment-like in texture. Each of the flat, ovoid capsules contains scores of baby whelks, incredible in the minute perfection of their shells. Sometimes a few remain in an egg string found on the beach; they rattle against the hard walls of the capsule like peas in a dried pod.

Perhaps the most baffling of all objects found on beaches are the egg cases of the sand collar snail or moon snail. If someone had cut a doll's shoulder cape out of a piece of fine sandpaper, the result would be about the same. The "collars" produced by the various species of the family of moon snails differ in size and, though slightly, in shape. In some the edges are smooth, in others scalloped. The arrangement of the eggs also follows slightly different patterns in the various species. This strange receptacle for the eggs of the snail is formed as a sheet of mucus pushed out from under the foot and molded on the outside of the shell. This results in the collar shape. The eggs are attached to the under side of the collar, which becomes completely impregnated with sand grains.

Mingled with the bits and fragments of sea creatures are the reminders of man's invasion of the sea—spars, pieces of rope, bottles, barrels, boxes of many shapes and sizes. If these have been long at sea, they bring their own collection of sea life, for in their period of drifting in the currents, they have served as a solid place of attachment for the searching larvae of the plankton.

On our Atlantic coast, the days following a northeast blow or a tropical storm are a time to look for the driftage of open ocean. I remember such a day on the beach at Nags Head, after a hurricane had passed by at sea during the night. The wind was still blowing a gale; there was a fine wild surf. That day the beach was strewn with many bits of driftwood, branches of trees, and heavy planks and spars, many of which bore growths of Lepas, the gooseneck barnacle of the open sea. One long plank was studded with tiny barnacles the size of a mouse's ear; on some of the other drifted timbers the barnacles had grown to a length of an inch or more, exclusive of the stalk. The size of the encrusting barnacles is a rough index of the time the spar has been at sea. In the profusion of their growth on almost every piece of timber one senses the incredible abundance of barnacle larvae drifting in the sea, ready to grasp any firm object adrift in their fluid world, for by strange irony none of them could complete their development in the sea water alone. Each of those weird-looking little beings, rowing through the water with feathered appendages, had to find a hard surface to which it could attach before assuming the adult form.

The life history of these stalked barnacles is very similar to that of the acorn barnacles of the rocks. Within the hard shells is a small crustacean body, bearing feathered appendages with which to sweep food into their mouths. The chief difference is that the shells are borne on a fleshy stalk instead of arising from a flat base firmly cemented to the substratum. When the animals are not feeding, the shells can be tightly closed, as in the rock barnacle; when they open to feed, there are the same sweeping, rhythmical motions of the appendages.

Seeing on the shore a branch from some tree that evidently has been long adrift and now is generously sprinkled with the fleshy brown stalks and the ivory-hued shells of the barnacle, with their marginal tints of blue and red, one can remember with tolerant understanding the old medieval misconception that conferred on these strange crustaceans the name "goose barnacle." The seventeenth-century English botanist John Gerard compiled a description of the "goose tree" or "Barnakle tree" on the basis of the following experience: "Traveling upon the shores of our English coast between Dover and Rummey, I founde the trunke of an olde rotten tree, which ... we drewe out of the water upon dry lande; on this rotten tree I founde growing many thousands of long crimson bladders ... at the neather end whereof did grow a shell fish, fashioned somewhat like a small Muskle ... which after I had opened ... I found living things that were very naked, in shape like a Birde; in others, the Birde covered with soft downe, the shell halfe open, and the Birde readie to fall out, which no doubt were the foules called Barnakles." Evidently Gerard's imaginative eye saw in the appendages of the barnacles the resemblance to a bird's feathers. On this slender basis he built the following pure fabrication: "They spawne as it were in March and April; the Geese are formed in Maie and June, and come to fulnesse of feathers in the moneth after." And so in many an old work of un-natural history from this time on, we see drawings of trees bearing fruit in the form of barnacles, and geese emerging from the shells to fly away.

Old spars and water-soaked timbers cast on the beach are full of the workings of the shipworm—long cylindrical tunnels penetrating all parts of the wood. Usually nothing remains of the creatures themselves except occasional fragments of their small calcareous shells; these proclaim that the shipworm is a true mollusk, despite its long, slender, and wormlike body.

There were shipworms long before there were men; yet within his own short tenancy of earth, man has greatly increased their numbers. The shipworm can live only in wood; if its young fail to discover some woody substance at a critical period of their existence, they die. This absolute dependence of a sea creature on something derived from the continents seems strange and incongruous. There could have been no shipworms until woody plants evolved on land. Their ancestors probably were clamlike forms burrowing in mud or clay, merely using their excavated holes as a base from which to extract the plankton of the sea. Then after trees evolved, these forerunners of the shipworms adapted themselves to a new habitat—the relatively few forest trees brought into the sea by rivers. But their numbers over all the earth must have been small until, scant thousands of years ago, men began to send wooden vessels across the sea and to build wharves at its edge; in all such wooden structures, the shipworm found a greatly extended range, to the cost of the human race.

The shipworm's place in history is secure. It was the scourge of the Romans with their galleys, of the seagoing Greeks arid Phoenicians, of the explorers of the New World. In the 1700's it riddled the dikes that the Dutch had built to keep out the sea; by so doing it threatened the very life of Holland. (As an academic by-product, the first extensive studies of the shipworm were made by Dutch scientists, to whom knowledge of its biology had become a matter of life and death. Snellius, in 1733, pointed out for the first time that this animal is a clamlike mollusk, not a worm.) About 1917 the shipworm invaded the harbor of San Francisco. Before its inroads were even suspected, ferry slips had begun to collapse, and wharves and loaded freight cars fell into the harbor. During the Second World War, especially in all tropical waters, the shipworm was an unseen but powerful enemy.

The female of the common shipworm retains the young in her burrow until they have attained the larval stage. Then they are launched into the sea—each a tiny being enclosed in two protective shells, looking like any other young bivalve. If it encounters wood when it has reached the threshold of adulthood, all goes well. It puts out a slender byssus thread as an anchor, a foot develops, and the shells become modified into efficient cutting tools, for rows of sharp ridges appear on their outer surfaces. The burrowing begins. With a powerful muscle, the animal scrapes the ridged shell against the wood, revolving meanwhile so that a smooth, cylindrical burrow is cut. As the burrow is extended, usually with the grain of the wood, the body of the shipworm grows. One end remains attached to the wall near the tiny point of entrance. This bears the siphons through which contact with the sea is maintained. The penetrating end carries the small shells. Between stretches a body that is thin as a lead pencil, but may reach a length of eighteen inches. Although a timber may be infested with hundreds of larvae, the burrows of the shipworms never interfere with each other. If an animal finds itself coming close to another burrow, it invariably turns aside. As it bores, it passes the loosened fragments of wood through its digestive tract. Some of the wood is digested and converted into glucose. This ability to digest cellulose is rare in the animal world—only certain snails, certain insects, and a very few others possess it. But the shipworm makes little use of this difficult art, and feeds chiefly on the rich plankton streaming through its body.

Other timbers on the beach bear the marks of the wood piddock. These are shallow holes that penetrate only the outer portions just beneath the bark, but they are broad and cleanly cylindrical. The boring piddock is seeking only shelter and protection. Unlike the shipworm, it does not digest the wood, but lives only on the plankton that it draws into its body through a protruding siphon.

Empty piddock holes sometimes attract other lodgers, as abandoned birds' nests may become homes for insects. On the muddy banks of salt creeks at Bears Bluff in South Carolina, I have picked up timbers riddled with holes. Once stout little white-shelled piddocks dwelt in them. The piddocks were long since dead and even the shells were gone, but in each hole was a dark glistening body like a raisin embedded in a cake. They were the contracted tissues of small anemones, finding there, in this world of silt-laden water and yielding mud, that bit of firm foundation which anemones must have. Seeing anemones in such an improbable place, one wonders how the larvae happened to be there, ready to seize the chance opportunity presented by that timber with its neatly excavated apartments; and one is struck anew by the enormous waste of life, remembering that for each of these anemones that succeeded in finding a home, many thousands must have failed.

Always, then, in this flotsam and jetsam of the tide lines, we are reminded that a strange and different world lies offshore. Though what we see here may be but the husks and fragments of life, through it we are made aware of life and death, of movement and change, of the transport of living things by ocean currents, by tides, by wind-driven waves. Some of these involuntary migrants are adults. They may perish in mid-journey; a few, being transported into a new home and finding there conditions that are favorable, may survive, may even produce surviving young to extend the range of the species. But many others are larvae, and whether or not they will make a successful landing depends on many things—on the length of their larval life (can they wait for a distant landfall before they reach the stage when they must take up an adult existence?)—on the temperature of the water they encounter—on the set of the currents that may carry them to favoring shoals, or off into deep water where they will be lost.

And so, walking the beach, we become aware of a most fascinating problem—the colonization of the shore, and especially of those "islands" of rock (or the semblance of rock) that occur in the midst of a sea of sand. For whenever a seawall is built, or a jetty, or pilings are sunk for a pier or a bridge, or rock, long hidden from sun and buried even beneath the sea, emerges again on the ocean floor, these hard surfaces immediately become peopled with typical animals of the rocks. But how did the colonizing rock fauna happen to be at hand—here in the midst of a sandy coast that stretches for hundreds of miles to north and south?

Pondering the answer, we become aware of that ceaseless migration, for the most part doomed to futility, yet ensuring that always, when opportunity arises, Life shall be waiting, ready to take advantage. For the ocean currents are not merely a movement of water; they are a stream of life, carrying always the eggs and young of countless sea creatures. They have carried the hardier ones across oceans, or step by step on long coastwise journeys. They have carried some along deep, hidden passageways where cold currents flow along the floor of the ocean. They have brought inhabitants to populate new islands pushing above the surface of the sea. These things they have done, we must suppose, since first there was life in the sea.

And as long as the currents move on their courses there is the possibility, the probability, even the certainty, that some particular form of life will extend its range, will come to occupy new territory.

As almost nothing else does, this to me expresses the pressure of the life force—the intense, blind, unconscious will to survive, to push on, to expand. It is one of life's mysteries that most of the participants in this cosmic migration are doomed to failure; it is no less mysterious that their failure turns into success when, for all the billions lost, a few succeed.

V
THE CORAL COAST

I DOUBT that anyone can travel the length of the Florida Keys without having communicated to his mind a sense of the uniqueness of this land of sky and water and scattered mangrove-covered islands. The atmosphere of the Keys is strongly and peculiarly their own. It may be that here, more than in most places, remembrance of the past and intimations of the future are linked with present reality. In bare and jaggedly corroded rock, sculptured with the patterns of the corals, there is the desolation of a dead past. In the multicolored sea gardens seen from a boat as one drifts above them, there is a tropical lushness and mystery, a throbbing sense of the pressure of life; in coral reef and mangrove swamp there are the dimly seen foreshadowings of the future.

This world of the Keys has no counterpart elsewhere in the United States, and indeed few coasts of the earth are like it. Offshore, living coral reefs fringe the island chain, while some of the Keys themselves are the dead remnants of an old reef whose builders lived and flourished in a warm sea perhaps a thousand years ago. This is a coast not formed of lifeless rock or sand, but created by the activities of living things which, though having bodies formed of protoplasm even as our own, are able to turn the substance of the sea into rock.

The living coral coasts of the world are confined to waters in which the temperature seldom falls below 70° F. (and never for prolonged periods), for the massive structures of the reefs can be built only where the coral animals are bathed by waters warm enough to favor the secretion of their calcareous skeletons. Reefs and all the associated structures of a coral coast are therefore restricted to the area bounded by the Tropics of Cancer and Capricorn. Moreover, they occur only on the eastern shores of continents, where currents of tropical water are carried toward the poles in a pattern determined by the earth's rotation and the direction of the winds. Western shores are inhospitable to corals because they are the site of upwellings of deep, cold water, with cold coastwise currents running toward the equator.

In North America, therefore, California and the Pacific coast of Mexico lack corals, while the West Indian region supports them in profusion. So do the coast of Brazil in South America, the tropical east African coast, and the northeastern shores of Australia, where the Great Barrier Reef creates a living wall for more than a thousand miles.

Within the United States the only coral coast is that of the Florida Keys. For nearly 200 miles these islands reach southwestward into tropical waters. They begin a little south of Miami where Sands, Elliott, and Old Rhodes Keys mark the entrance to Biscayne Bay; then other islands continue to the southwest, skirting the tip of the Florida mainland, from which they are separated by Florida Bay, and finally swinging out from the land to form a slender dividing line between the Gulf of Mexico and the Straits of Florida, through which the Gulf Stream pours its indigo flood.

To seaward of the Keys there is a shallow area three to seven miles wide where the sea bottom forms a gently sloping platform under depths generally less than five fathoms. An irregular channel (Hawk Channel) with depths to ten fathoms traverses these shallows and is navigable by small boats. A wall of living coral reefs forms the seaward boundary of the reef platform, standing on the edge of the deeper sea (see page 198).

The Keys are divided into two groups that have a dual nature and origin. The eastern islands, swinging in their smooth arc 110 miles from Sands to Loggerhead Key, are the exposed remnants of a Pleistocene coral reef. Its builders lived and flourished in a warm sea just before the last of the glacial periods, but today the corals, or all that remains of them, are dry land. These eastern Keys are long, narrow islands covered with low trees and shrubs, bordered with coral limestone where they are exposed to the open sea, passing into the shallow waters of Florida Bay through a maze of mangrove swamps on the sheltered side. The western group, known as the Pine Islands, are a different kind of land, formed of limestone rock that had its origin on the bottom of a shallow interglacial sea, and is now raised only slightly above the surface of the water. But in all the Keys, whether built by the coral animals or formed of solidifying sea drift, the shaping hand is the hand of the sea.

In its being and its meaning, this coast represents not merely an uneasy equilibrium of land and water masses; it is eloquent of a continuing change now actually in progress, a change being brought about by the life processes of living things. Perhaps the sense of this comes most clearly to one standing on a bridge between the Keys, looking out over miles of water, dotted with mangrove-covered islands to the horizon. This may seem a dreamy land, steeped in its past. But under the bridge a green mangrove seedling floats, long and slender, one end already beginning to show the development of roots, beginning to reach down through the water, ready to grasp and to root firmly in any muddy shoal that may lie across its path. Over the years the mangroves bridge the water gaps between the islands; they extend the mainland; they create new islands. And the currents that stream under the bridge, carrying the mangrove seedling, are one with the currents that carry plankton to the coral animals building the offshore reef, creating a wall of rocklike solidity, a wall that one day may be added to the mainland. So this coast is built.

To understand the living present, and the promise of the future, it is necessary to remember the past. During the Pleistocene, the earth experienced at least four glacial stages, when severe climates prevailed and immense sheets of ice crept southward. During each of these stages, large volumes of the earth's water were frozen into ice, and sea level dropped all over the world. The glacial intervals were separated by milder interglacial stages when, with water from melting glaciers returning to the sea, the level of the world ocean rose again. Since the most recent Ice Age, known as the Wisconsin, the general trend of the earth's climate has been toward a gradual, though not uniform warming up. The interglacial stage preceding the Wisconsin glaciation is known as the Sangamon, and with it the history of the Florida Keys is intimately linked.

The corals that now form the substance of the eastern Keys built their reef during that Sangamon interglacial period, probably only a few tens of thousands of years ago. Then the sea stood perhaps 100 feet higher than it does today, and covered all of the southern part of the Florida plateau. In the warm sea off the sloping southeastern edge of that plateau the corals began to grow, in water somewhat more than 100 feet deep. Later the sea level dropped about 30 feet (this was in the early stages of a new glaciation, when water drawn from the sea was falling as snow in the far north); then another 30 feet. In this shallower water the corals flourished even more luxuriantly and the reef grew upward, its structure mounting close to the sea surface. But the dropping sea level that at first favored the growth of the reef was to be its destruction, for as the ice increased in the north in the Wisconsin glacial stage, the ocean level fell so low that the reef was exposed and all its living coral animals were killed. Once again in its history the reef was submerged for a brief period, but this could not bring back the life that had created it. Later it emerged again and has remained above water, except for the lower portions, which now form the passes between the Keys. Where the old reef lies exposed, it is deeply corroded and dissected by the dissolving action of rain and the beating of salt spray; in many places the old coral heads are revealed, so distinctly that the species are identifiable.

While the reef was a living thing, being built up in that Sangamon sea, the sediments that have more recently become the limestone of the western group of Keys were accumulating on the landward side of the reef. Then the nearest land lay 150 miles to the north, for all the southern end of the present Florida peninsula was submerged. The remains of many sea creatures, the solution of limestone rocks, and chemical reactions in the sea water contributed to the soft ooze that covered the shallow bottoms. With the changing sea levels that followed, this ooze became compacted and solidified into a white, fine-textured limestone, containing many small spherules of calcium carbonate resembling the roe of fish; because of this characteristic it is sometimes known as "oolitic limestone," or "Miami oolite." This is the rock immediately underlying the southern part of the Florida mainland. It forms the bed of Florida Bay under the layer of recent sediments, and then rises above the surface in the Pine Islands, or western Keys, from Big Pine Key to Key West. On the mainland, the cities of Palm Beach, Fort Lauderdale, and Miami stand on a ridge of this limestone formed when currents swept past an old shore line of the peninsula, molding the soft oozes into a curving bar. The Miami oolite is exposed on the floor of the Everglades as rock of strangely uneven surface, here rising in sharp peaks, there dropping away in solution holes. Builders of the Tamiami Trail and of the highway from Miami to Key Largo dredged up this limestone along the rights of way and with it built the foundations on which these highways are laid.

Knowing this past, we can see in the present a repetition of the pattern, a recurrence of earth processes of an earlier day. Now, as then, living reefs are building up offshore; sediments are accumulating in shallow waters; and the level of the sea, almost imperceptibly but certainly, is changing.

Off this coral coast the sea lies green in the shallows, blue in the far distances. After a storm, or even after a prolonged southeasterly blow, comes "white water." Then a thick, milk-white, richly calcareous sediment is washed out of the reefs and stirred from its deep beds over the floor of the reef flat. On such days the diving mask and the aqualung may as well be left behind, for the underwater visibility is little better than in a London fog.

"White water" is the indirect result of the very high rate of sedimentation that prevails in the shallows around the Keys. Anyone who wades out even a few steps from the shore notices the white, siltlike substance adrift in the water and accumulating on the bottom. It has visibly rained down on every surface. Its fine dust lies over sponge and gorgonian and anemone; it chokes and buries the low-growing algae and lies whitely over the dark bulks of the big loggerhead sponges. The wader stirs up clouds of it; winds and strong currents set it in motion. Its accumulation is going on at an astonishing rate; sometimes, after a storm, two or three inches of new sediment are deposited from one high tide to the next. It comes from various sources. Some is mechanically derived from the disintegration of dead plants and animals—mollusk shells, lime-depositing algae, coral skeletons, tubes of worms or snails, spicules of gorgonians and sponges, skeletal plates of holothurians. It is also derived in part from chemical precipitation of the calcium carbonate present in the water. This, in turn, has been leached out of the vast expanses of limestone rock that compose the surface of southern Florida, and has been carried to the sea by rivers and by the slow drainage of the Everglades.

A few miles outside the chain of the present Keys is the reef 5 of living coral, forming the seaward rim of the shallows, and overlooking a steep descent into the trough of the Florida straits. The reefs extend from Fowey Rocks, south of Miami, to the Marquesas and Tortugas and in general they mark the 10-fathom depth contour. But often they rise to lesser depths and here and there they break the surface as tiny offshore islands, many of them marked by lighthouses.

Drifting over the reef in a small boat and peering down through a glass-bottomed bucket, one finds it hard to visualize the whole terrain, for so little of it can be seen at a time. Even a diver exploring more intimately finds it difficult to realize he is on the crest of a high hill, swept by currents instead of winds, where gorgonians are the shrubbery and stands of elkhorn coral are trees of stone. Toward the land, the sea floor slopes gently down from this hilltop into the wide water-filled valley of Hawk Channel; then it rises again and breaks water as a chain of low-lying islands—the Keys. But on the seaward side of the reef the bottom descends quickly into blue depths. Live corals grow down to a depth of about 10 fathoms. Below that it is too dark, perhaps, or there is too much sediment, and instead of living coral there is a foundation of dead reef, formed at some time when the sea level was lower than it is today. Out where the water is about 100 fathoms deep there is a clean rock bottom, the Pourtal[[[grave. gif]]]s Plateau; its fauna is rich, but the corals that live here are not reef builders. Between 300 and 500 fathoms sediments have again accumulated on a slope that descends to the trough of the Florida straits—the channel of the Gulf Stream.

As for the reef itself, many thousand thousand beings—plant and animal, living and dead—have entered into its composition. Corals of many species, building their little cups of lime and with them fashioning many strange and beautiful forms, are the foundation of the reef. But besides the corals there are other builders and all the interstices of the reef are filled with their shells or their limy tubes, or with coral rock cemented together with building stones of the most diverse origin. There are colonies of tube-building worms and there are mollusks of the snail tribe whose contorted, tubular shells may be intertwined into massive structures. Calcareous algae, which have the property of depositing lime in their living tissues, form part of the reef itself or, growing abundantly over the shallows on the landward side, add their substance at death to the coral sand of which limestone rock is later formed. The horny corals or gorgonians, known as sea fans and sea whips, all contain limestone spicules in their soft tissues. These, along with lime from starfish and sea urchins and sponges and an immense number of smaller creatures, will eventually, with the passage of time and through the chemistry of the sea, come to form part of the reef.

Along with those that build are others that destroy. The sulphur sponge dissolves away the calcareous rock. Boring mollusks riddle it with their tunnels, and worms with sharp, biting jaws eat into it, weakening its structure and so hastening the day when a mass of coral will yield to the force of the waves, will break away, and perhaps roll down the seaward face of the reef into deeper water.

The basis of this whole complex association is a minute creature of deceptively simple appearance, the coral polyp. The coral animal is formed on the same general lines as the sea anemone. It is a double-walled tube of cylindrical shape, closed at the base and open at the free end, where a crown of tentacles surrounds the mouth. The important difference—the fact on which the existence of coral reefs depends—is this: the coral polyp has the ability to secrete lime, forming a hard cup about itself. This is done by cells of the outer layer, much as the shell of a mollusk is secreted by an outer layer of soft tissue—the mantle. So the anemone-like coral polyp comes to sit in a compartment formed of a substance as hard as rock. Because the "skin" of the polyp is turned inward at intervals in a series of vertical folds, and because all of this skin is actively secreting lime, the cup does not have a smooth circumference, but is marked by partitions projecting inward, forming the starlike or flowerlike pattern familiar to anyone who has examined a coral skeleton.

Most corals build colonies of many individuals. All the individuals of any one colony, however, are derived from a single fertilized ovum that matured and then began to form new polyps by budding. The colony has a shape characteristic of the species—branched, boulderlike, flatly encrusting, or cup-shaped. Its core is solid, for only the surface is occupied by living polyps, which may be widely separated in some species or closely crowded in others. It is often true that the larger and more massive the colony, the smaller the individuals that compose it; the polyps of a branching coral taller than a man may themselves be only an eighth of an inch high.

The hard substance of the coral colony is usually white, but may take on the colors of minute plant cells that live within the soft tissues in a relation of mutual benefit. There is the exchange usual in such relations, the plants getting carbon dioxide and the animals making use of the oxygen given off by the plants. This particular association may have a deeper significance, however. The yellow, green, or brown pigments of the algae belong to the group of chemical substances known as carotinoids. Recent studies suggest that these pigments in the imprisoned algae may act on the corals, serving as "internal correlators" to influence the processes of reproduction. Under normal conditions, the presence of the algae seems to benefit the coral, but in dim light the coral animals rid themselves of the algae by excreting them. Perhaps this means that in weak light or in darkness the whole physiology of the plant is changed and the products of its metabolism are altered to something harmful, so that the animal must expel the plant guest.

Within the coral community there are other strange associations. In the Florida Keys and elsewhere in the West Indian region, a gall crab makes an oven-shaped cavity on the upper surface of a colony of living brain coral. As the coral grows the crab manages to keep open a semicircular entrance through which, while young, it enters and leaves its den. Once full grown, however, the crab is believed to be imprisoned within the coral. Few details of the existence of this Florida gall crab are known, but in a related species in corals of the Great Barrier Reef only the females form galls. The males are minute, and apparently visit the females in the cavities where they are imprisoned. The female of this species depends on straining food organisms from indrawn currents of sea water and its digestive apparatus and appendages are much modified.

Everywhere, throughout the whole structure of the reef as well as inshore, the horny corals or gorgonians are abundant, sometimes outnumbering the corals. The violet-hued sea fan spreads its lace to the passing currents, and from all the structure of the fan innumerable mouths protrude through tiny pores, and tentacles reach out into the water to capture food. The little snail known as the flamingo tongue, wearing a solid and highly polished shell, often lives on the sea fans. The soft mantle, extended to cover the shell, is a pale flesh color with numerous black, roughly triangular markings. The gorgonians known as sea whips are more abundant, forming dense stands of undersea shrubbery, often waist-high, sometimes as tall as a man. Lilac, purple, yellow, orange, brown, and buff are the colors worn by these gorgonians of the coral reefs.

Encrusting sponges spread their mats of yellow, green, purple, and red over the walls of the reef; exotic mollusks like the jewel box and the spiny oyster cling to it; long-spined sea urchins make dark, bristling patches in the hollows and crevices; and schools of brightly colored fishes twinkle along the façade of the reef where the lone hunters, the gray snapper and the barracuda, wait to seize them.

At night the reef comes alive. From every stony branch and tower and domed façade, the little coral animals, who, avoiding daylight, had remained shrunken within their protective cups until darkness fell, now thrust out their tentacled heads and feed on the plankton that is rising toward the surface. Small crustacea and many other forms of microplankton, drifting or swimming against a branch of coral, are instant victims of the myriad stinging cells with which each tentacle is armed. Minute though the individual plankton animals be, the chances of passing unharmed through the interlacing branches of a stand of elkhorn coral seem slender indeed.

Other creatures of the reef respond to night and darkness and many of them emerge from the grottoes and crevices that served as daytime shelter. Even that strange hidden fauna of the massive sponges—the small shrimps and amphipods and other animals that live as unbidden guests deep within the canals of the sponge—at night creep up along those dark and narrow galleries and collect near their thresholds as though looking out upon the world of the reef.

On certain nights of the year, extraordinary events occur over the reefs. The famed palolo worm of the South Pacific, moved to gather in its prodigious spawning swarms on a certain moon of a certain month—and then only—has its less-known counterpart in a related worm that lives in the reefs of the West Indies and at least locally in the Florida Keys. The spawning of this Atlantic palolo has been observed repeatedly about the Dry Tortugas reefs, at Cape Florida, and in several West Indian localities. At Tortugas it takes place always in July, usually when the moon reaches its third quarter, though less often on the first quarter. The worms never spawn on the new moon.

The palolo inhabits burrows in dead coral rock, sometimes appropriating the tunnelings of other creatures, sometimes excavating its burrow by biting away fragments of rock. The life of this strange little creature seems to be ruled by light. In its immaturity the palolo is repelled by light—by sunlight, by the light of the full moon, even by paler moonlight. Only in the darkest hours of the night, when this strong inhibition of the light rays is removed, does it venture from its burrow, creeping out a few inches in order to nibble at the vegetation on the rocks. Then, as the season for spawning approaches, remarkable changes take place within the bodies of the worms. With the maturing of the sex cells, the segments of the posterior third of each animal take on a new color, deep pink in the males, greenish gray in the females. Moreover, this part of the body, distended with eggs or sperm, becomes exceedingly thin-walled and fragile, and a noticeable constriction develops between this and the anterior part of the worm.

At last there comes a night when these worms—so changed in their physical beings—respond in a new way to the light of the moon. No longer does the light repel and hold them prisoners within their burrows. Instead, it draws them out to the performance of a strange ritual. The worms back out of their burrows, thrusting out the swollen, thin-walled posterior ends, which immediately begin a series of twisting movements, writhing in spiral motions until suddenly the body breaks at the weak point and each worm becomes two. The two parts have different destinies—the one to remain behind in the burrow and resume the life of the timid forager of the dark hours, the other to swim up toward the surface of the sea, to become one of a vast swarm of thousands upon thousands of worms joining in the spawning activities of the species.

During the last hours of the night the number of swarming worms increases rapidly, and when dawn comes the sea over the reef is almost literally filled with them. When the first rays of the sun appear, the worms, strongly stimulated by the light, begin to twist and contract violently, their thin-walled bodies burst open, and the eggs from some and sperm from others are cast into the sea. The spent and empty worms may continue to swim weakly for a short time, preyed upon by fish that gather for a feast, but soon all that remain have sunk to the bottom and died. But floating at the surface of the sea are the fertilized eggs, drifting over areas many feet deep and acres in extent. Within them swift changes have begun—the division of cells, the differentiation of structure. By evening of that same day the eggs have yielded up tiny larvae, swimming with spiral motions through the sea. For about three days the larvae live at the surface; then they become burrowers in the reefs below until, a year hence, they will repeat the spawning behavior of their kind.

Some related worms that swarm periodically about the Keys and the West Indies are luminous, creating beautiful pyrotechnic displays on dark nights. Some people believe that the mysterious light reported by Columbus as seen by him on the night of October 11, "about four hours before making the landfall and an hour before moonrise," may have been a display of some of these "fireworms."

The tides pouring in from the reefs and sweeping over the flats come to rest against the elevated coral rock of the shore. On some of the Keys the rock is smoothly weathered, with flattened surfaces and rounded contours, but on many others the erosive action of the sea has produced a rough and deeply pitted surface, reflecting the solvent action of centuries of waves and driven salt spray. It is almost like a stormy sea frozen into solidity, or as the surface of the moon might be.

Little caves and solution holes extend above and below the line of the high tide. In such a place I am always strongly aware of the old, dead reef beneath my feet, and of the corals whose patterns, now crumbling and blurred, were once the delicately sculptured vessels that held the living creatures. All the builders now are dead—they have been dead for thousands of years—but that which they created remains, a part of the living present.

Crouching on the jagged rocks, I hear little murmurings and whisperings born of the movements of air and water over these surfaces—the audible voice of this nonhuman, intertidal world. There are few obvious signs of life to break the spell of brooding desolation. Perhaps a dark-bodied isopod—a sea roach—darts across the dry rock to disappear into one of the small sea caves, daring exposure to light and to sharp-eyed enemies only for the moment of its swift passage from one dark recess to another. There are thousands of its kind in the coral rock, but not until darkness covers the shore will they come out in numbers to search for the bits of animal and vegetable refuse that are their food.

At the high-tide line, growths of microscopic plants darken the coral rock, tracing that mysterious black line that marks the sea's edge on all rocky coasts of the world. Because of the irregular surface and deep dissection of the coral rock, the sea runs in under the high-tide rocks by way of crevices and depressions, and so the black zone darkens the jagged peaks and the rims of holes and little caves, while lighter rock of a yellowish-gray hue lines the depressions below that controlling tidal level.

Small snails whose shells are boldly striped or checked in black and white—the neritas—crowd down into cracks and cavities in the coral or rest on open rock surfaces waiting for the return of the tide when they can feed. Others, in rounded shells with roughly beaded surfaces, belong to the periwinkle tribe. Like many others of their kind, these beaded periwinkles are making a tentative invasion of the land, living under rocks or logs high on the shore or even entering the fringe of land vegetation. Black horn shells live in numbers just below the line of the high tides, feeding on the algal film over the rocks. The living snails are held by some intangible bonds to this tidal level, but the shells discarded after their death are found and taken as habitations by the smallest of the hermit crabs, who then carry them down onto the lower levels of the shore.

These deeply eroded rocks are the home of the chitons, whose primitive form harks back to some ancient group of mollusks of which they are the only living representatives. Their oval bodies, covered with a jointed shell of eight transverse plates, fit into depressions in the rocks when the tide is out. They grip the rocks so strongly that even heavy waves can get no hold on their sloping contours. When the high tide covers them, they begin to creep about, resuming their rasping of vegetation from the rocks, their bodies swaying to and fro in time to the scraping motions of the radula or file-like tongue. Month in and month out, a chiton moves only a few feet in any direction; because of this sedentary habit, the spores of algae and the larvae of barnacles and tube-building worms settle upon its shell and develop there. Sometimes, in dark wet caves, the chitons pile up, one on top of another, and each scrapes algae off the back of the one beneath it. In a small way these primitive mollusks may be an agent of geologic change as they feed on the rocks, each removing, along with the algae, minute scrapings of rock particles and so, over the centuries and the millennia in which this ancient race of beings has lived its simple life, contributing to the processes of erosion by which earth surfaces are worn away.

On some of these Keys a small intertidal mollusk called Onchidium lives deep in little rock caverns, the entrances of which are often overgrown by colonies of mussels. Although it is a mollusk and a snail, Onchidium has no shell. It belongs to a group that consists largely of land snails or slugs, in many of which the shell is lacking or concealed. Onchidium inhabits tropical seashores, living usually on beaches of roughly eroded rock. As the tide falls, processions of small black slugs emerge from their doorways, wriggling and pushing their way out through the impeding mussel threads, a dozen or more individuals coming out of a common cave to feed on the rocks, from which they scrape vegetation as the chitons do. As they emerge, each is invested with a tunic of slime that makes it look jet black, wet, and shining; in wind and sun the little slug dries to a deep blue-black, over which is a slight, milky bloom.

On these journeys the slugs seem to follow haphazard or irregular paths over the rocks. They continue feeding as the tide falls to its lowest ebb, and even as it turns and begins to rise. About half an hour before the returning sea has reached them, and before so much as a drop of water has splashed into their nests, all of the slugs cease their grazing and begin to return to the home nest. While the outgoing path was meandering, the return is by a direct route. The members of each community return to their own nest, even though the way may lie over greatly eroded rock surfaces and even though the path may cross the routes of other slugs returning to other nests. All of the individuals belonging to one nest-community, even though they may have been widely separated while feeding, begin the return journey at almost the same moment. What is the stimulus? It is not the returning water, for that has not touched them; when it laps again over their rocks they will be safe within their nests.

The whole pattern of behavior of this little creature is puzzling. Why should it be drawn to live again at the edge of the sea that its ancestors deserted thousands or millions of years ago? It comes forth only when the tide has fallen, then, somehow sensing the impending return of the sea and seeming to remember its recent affinities with the land, it hurries to safety before the tide can find it and carry it away. How has it acquired this behavior, attracted yet repelled by the sea? We can only ask these questions; we cannot answer them.

For its protection during the feeding journeys, Onchidium is equipped with means of detecting and driving away its enemies. Minute papillae on its back are sensitive to light and passing shadows. Other, stouter papillae associated with the mantle are equipped with glands that secrete a milky, highly acid fluid. If the animal is suddenly disturbed, it expels spurting streams of this acid, the streams breaking up in the air to a fine spray that may be thrown five or six inches, or as much as a dozen times the length of the animal. The old German zoologist Semper, who studied a species of Onchidium in the

Philippines, believed this dual equipment served to protect the slug from the beach-hopping blenny, a fish of many tropical mangrove coasts that leaps along above the tide, feeding on Onchidium and crabs. Semper thought the slugs could detect the shadow of an approaching fish and drive off the enemy by discharging the white acid spray. In Florida or elsewhere in the West Indian region there is no fish that comes out of water to pursue its prey. On the rocks where Onchidium must feed there are, however, scrambling crabs and isopods whose jostlings might well push the slugs into the water, for they have no means of gripping the rocks. For whatever reason, the slugs react to the crabs and to the isopods as to dangerous enemies, responding to their touch by discharging the repellant chemical.

In the strip between tropical tide lines, conditions are difficult for nearly all forms of life. The heat of the sun increases the hazards of exposure during the withdrawal of the tide. The shifting layers of choking sediment, accumulating on flat or gently sloping surfaces, discourage many plants and animals of types that inhabit rocky shores in the clearer, cooler waters of the north. Instead of the vast barnacle and mussel fields of New England there are only scattering patches of these creatures, varying from Key to Key but never really abundant. Instead of the great rockweed forests of the north, there are only scattered growths of small algae, including various brittle, lime-secreting forms, none of which offer shelter or security to any considerable number of animals.

If the area marked out by the advance and retreat of the neap tides is in general inhospitable, there are nevertheless two forms of life—one plant, one animal—that are thoroughly at home there, and live in profusion nowhere else. The plant is a peculiarly-beautiful alga that resembles spheres of green glass clustered together in irregular masses. It is Valonia, the sea bottle, a green alga that forms large vesicles filled with a sap that bears a definite chemical relation to the water about it, varying the proportions of its contained ions of sodium and potassium according to variations in the intensity of sunlight, the exposure to surf, and other conditions of its world. Under overhanging rock and in other sheltered places it forms sheets and masses of its emerald globules, lying half buried in deep drifts of sediment.

The animal symbol of this intertidal world of coral is a group of snails whose whole structure and being represent an extraordinary contrast to the way of life typical of this class of mollusks. They are called the vermetid or "wormlike" snails. The shell is no ordinary gastropod spire or cone, but a loose uncoiled tube very like the calcareous tubes built by many worms. The species that inhabit this intertidal zone have become colonial, and their tubes form closely packed and intertwined masses.

The very nature of these vermetid snails and their departure from the form and habits of related mollusks are eloquent of the circumstances of their world and of the readiness of life to adapt itself to a vacant niche. Here on this coral platform the tide ebbs and flows twice daily, and each flood brings renewed food supplies from offshore. There is but one perfect way to exploit such rich supplies: to remain in one place and fish the currents as they stream by. This is done on other shores by such animals as the barnacles, the mussels, and the tube-building worms. It is not ordinarily a snail's way of life, but in adaptation these extraordinary snails have become sedentary, abandoning the typical roaming habit. No longer solitary, they have become gregarious to an extreme degree, living in crowded colonies, with shells so intertwined that early geologists called their formations "worm rock." And they have given up the snail habits of scraping food from the rocks or of hunting and devouring other animals of large size; instead they draw the sea water into their bodies and strain out the minute planktonic food organisms. The tips of their gills are thrust out and drawn through the water like nets—an adaptation probably unique in all the group of snail-like mollusks. The vermetids give their own clear demonstration of the plasticity of the living organism and its responsiveness to the world about it. Again and again, in group after group of widely different and unrelated animals, the same problem has been met and solved by the evolution of diverse structures that function for a common purpose. So the legions of the barnacles sweep food from the tides on a New England shore, using a modification of what in their relatives would be a swimming appendage; mole crabs gather by the thousand where surf sweeps the southern beaches, straining out food with the bristles of their antennae; and here on this coral shore the crowded masses of this strange snail filter the waters of the incoming tide through their gills. By becoming the imperfect, the atypical snail, they have become the perfectly adapted exploiter of the opportunities of their world.

The edge of the low tide is a dark line traced by colonies of short-spined, rock-boring sea urchins. Every hole and every depression in the coral rock bristles with their small dark bodies. One spot in the Keys lives in my memory as an urchin paradise. This is the seaward shore of one of the eastern group of islands, where the rock drops in an abrupt terrace, somewhat undercut and deeply eroded into holes and small caves, many with their roofs open to the sky. I have stood on the dry rock above the tide and looked down into these little water-floored, rock-walled grottoes, finding twenty-five to thirty urchins in one of these caverns that was no larger than a bushel basket. The caves shine with a green water-light in the sun, and in this light the globular bodies of the urchins have a reddish color of glowing, luminous quality, in rich contrast to the black spines.

A little beyond this spot the sea bottom slopes under water more gradually, with no undercutting. Here the rock borers seem to have taken over every niche that can afford shelter; they give the illusion of shadows beside each small irregularity of bottom. It is not certain whether they use the five short stout teeth on their under surfaces to scrape out holes in the rock, or perhaps merely take advantage of natural depressions to find a safe anchorage against the occasional storms that sweep this coast. For some inscrutable reason, these rock-boring urchins and related species in other parts of the world are bound to this particular tidal level, linked to it precisely and mysteriously by invisible ties that prevent their wandering farther out over the reef flat, although other species of urchins are abundant there.

Above and below the zone of the rock-boring urchins, closely crowded throngs of pale brown tubular creatures push up through the chalky sediment. When the tide leaves them their tissues retract and all that proclaims them to be animals is hidden; then one might pass them by as some strange marine fungi. With the return of the water their animal nature is revealed, and from each fawn-colored tube a crown of tentacles, of purest emerald green, is unfolded as each of these anemone-like creatures begins to search the tide for the food it has brought. Living where their very existence depends on keeping the delicate tissues of the tentacles above the choking dust of sediment, these zoanthids are able to stretch their bodies into slender threads where the sediments are deep, though normally their tubes are short and stout.

On the seaward side of many of the Keys the bottom slopes gently, with wading depths for perhaps a quarter of a mile or more. Once beyond the rock-boring sea urchins, the vermetid snails, and the green and brown jewel anemones, the bottom of coarse sand and coral fragments begins to be marked by dark patches of turtle grass, and larger animals begin to inhabit the reef flats. Sponges, dark and bulky, grow in water only deep enough to cover their massive forms. Small, shallow-water corals, somehow able to survive the rain of sediments that would be fatal to the larger reef-builders, erect their hard structures, stoutly branched or domed, on the floor of coral rock. The gorgonians, plantlike in their habit of growth, are a low shrubbery of delicate rose and brown and purple hues. And within and among and beneath them all is the infinitely varied fauna of a tropical coast, as many creatures that wander freely through the waters of this warm sea crawl or swim or glide over the flats.

Massive and inert, the loggerhead sponges by their appearance suggest nothing of the activity that goes on within their dark bulks. There is no sign of life for the casual passer-by to read, although if he waited and watched long enough he might sometimes see the deliberate closing of some of the round openings, large enough to admit an exploring finger, that penetrate the flat upper surface. These and other openings are the key to the nature of the giant sponge which, like even the smallest of its group, can exist only as long as it can keep the waters of the sea circulating through its body. Its vertical walls are pierced by intake canals of small diameter, groups of them covered by sieve plates with numerous perforations. From these the canals lead almost horizontally into the interior of the sponge, branching and rebranching into tubes of progressively smaller bore, to penetrate all the massive bulk of the sponge and finally to lead up to the large exit canals. Perhaps these exit holes are kept free of choking sediment by the strength of the outbound currents; at any rate they are the only part of the sponge that shows a pure black color, for the flour-like whiteness of the reef sediments has been sifted over all the sooty black surface of the body.

In its passage through the sponge, the water leaves a coating of minute food organisms and organic detritus on the walls of the canals; the cells of the sponge pick up the food, pass the digestible materials along from cell to cell, and return waste material to the flowing currents. Oxygen passes into the sponge cells; carbon dioxide is given off. And sometimes small sponge larvae, having undergone the early stages of their development within the parent sponge, detach themselves and enter the dark flowing river, to pass with it into the sea.

The intricate passageways, the shelter and available food they offer, have attracted many small creatures to live within the sponge. Some come and go; others never leave the sponge once they have taken up residence within it. One such permanent lodger is a small shrimp—one of the group known as snapping shrimp because of the sound made by snapping the large claw. Although the adults are imprisoned, the young shrimp, hatched from eggs adhering to the appendages of their mothers, pass out with the water currents into the sea and live for a time in the currents and tides, drifting, swimming, perhaps carried far afield. By mischance they may occasionally find their way into deep water where no sponges grow. But many of the young shrimp will in time find and approach the dark bulk of some loggerhead sponge and, entering it, will take up the strange life of their parents. Wandering through its dark halls, they scrape food from the walls of the sponge. As they creep along these cylindrical passageways, they carry their antennae and their large claws extended before them, as though to sense the approach of a larger and possibly dangerous creature, for the sponge has many lodgers of many species—other shrimps, amphipods, worms, isopods—and their numbers may reach into the thousands if the sponge is large.

There, on the flats off some of the Keys, I have opened small loggerheads and heard the warning snapping of claws as the resident shrimps, small, amber-colored beings, hurried into the deeper cavities. I had heard the same sound filling the air about me, as, on an evening low tide, I waded in to the shore. From all the exposed reef rock there were strange little knockings and hammerings, yet the sounds, to a maddening degree, were impossible to locate. Surely this nearby hammering came from this particular bit of rock; yet when I knelt to examine it closely there was silence; then from all around, from everywhere but this bit of rock at hand, all the elfin hammering was resumed. I could never find the little shrimps in the rocks, yet I knew they were related to those I had seen in the loggerhead sponges. Each has one immense hammer claw almost as long as the rest of its body. The movable finger of the claw bears a peg that fits into a socket in the rigid finger. Apparently the movable finger, when raised, is held in position by suction. To lower it, extra muscular force must be applied, and when the suction is overcome, it snaps into place with audible sound, at the same time ejecting a spurt of water from its socket. Perhaps the water jet repels enemies and aids in capturing prey, which may also be stunned by a blow from the forcibly retracted claw. Whatever the value of the mechanism, the snapping shrimps are so abundant in the shallows of tropical and subtropical regions, and snap their claws so incessantly, that they are responsible for much of the extraneous noise picked up on underwater listening devices, filling the water world with a continuous sizzling, crackling sound.

It was on the reef flats off Ohio Key, on a day early in May, that I had my first, startled encounter with tropical sea hares. I was wading over a part of the flat that had an unusually heavy growth of rather tall seaweeds when sudden movement drew my eyes to several heavy-bodied, foot-long animals moving among the weeds. They were a pale tan color, marked with black rings, and when I touched one cautiously with my foot, it responded instantly by expelling a concealing cloud of fluid the color of cranberry juice.

I had met my first sea hare years before on the North Carolina coast. It was a small creature about as long as my little finger, browsing peacefully among some seaweeds near a stone jetty. I slipped my hand under it and gently brought it toward me, then, its identity confirmed, I returned the little creature carefully to the algae, where it resumed its grazing. Only by drastic revision of my mental image could I accept these tropical creatures, which seemed to belong in some book of mythology, as relatives of that first little elfin being.

The large West Indian sea hares inhabit the Florida Keys as well as the Bahamas, Bermuda, and the Cape Verde Islands. Within their range they usually live offshore, but at the spawning season they move in to the shallows, where I had found them, to attach their eggs, in tangled threads, to the weeds near the low-tide mark. They are marine snails of a sort, but have lost their external shells and possess only an internal remnant, hidden by the soft mantle tissue. Two prominent tentacles suggestive of ears, and the rabbit-like body shape, are responsible for the common name (see page 235).

Whether because of its strange appearance, or because of its defensive fluids, often thought to be poisonous, the Old World sea hare has long had a secure place in folk lore, superstition, and witchcraft. Pliny declared it was poisonous to the touch, and recommended as an antidote asses' milk and ground asses' bones, boiled together. Apuleius, known chiefly as the author of The Golden Ass, became curious about the internal anatomy of the sea hare and persuaded two fishermen to bring him a specimen; whereupon he was accused of witchcraft and poisoning. Some fifteen centuries were to pass before anyone else ventured to publish a description of the internal anatomy of the creature—then Redi in 1684 described it, and although popular belief called it sometimes a worm, sometimes a holothurian, sometimes a fish, he placed it correctly, at least as to general relationships, as a marine slug. For the past century or more the harmless nature of the sea hares has been recognized for the most part, but although they are fairly well known in Europe and Great Britain, the American sea hares, largely confined to tropical waters, are less familiar animals.

Perhaps this anonymity is due in part to the infrequency of their spawning migrations into tidal waters. An individual animal is both male and female; it may function as either sex, or as both. In laying its eggs, the sea hare extrudes a long thread in little spurts, about an inch at a time, continuing the slow process until the string has reached a length sometimes as great as 65 feet, and contains about 100,000 eggs. As the pink or orange-colored thread is expelled it curls about the surrounding vegetation, forming a tangled mass of spawn. The eggs and the resulting young meet the common fate of marine creatures; many eggs are destroyed, being eaten by crustacea or other predators (even by their own kind), and many of the hatching larvae fail to survive the dangers of life in the plankton. In the drift of the currents the larvae are carried offshore, and when they undergo metamorphosis to the adult form and seek the bottom they are in deep water. Their color changes with changing food as they migrate shoreward: first they are a deep rose color, then they are brown, then olive-green like the adults. For one of the European species, at least, the known life history suggests a curious parallel with that of the Pacific salmon. With maturity, the sea hares turn shoreward to spawn. It is a journey from which there is no return; they do not reappear on the offshore feeding grounds, but apparently die after this single spawning.

The world of the reef flats is inhabited by echinoderms of every sort: starfishes, brittle stars, sea urchins, sand dollars, and holothurians all are at home on the coral rock, in the shifting coral sands, among the gorgonian sea gardens and the grass-carpeted bottoms. All are important in the economy of the marine world—as links in the living chains by which materials are taken from the sea, passed from one to another, returned to the sea, borrowed again. Some are important also in the geologic processes of earth building and earth destruction—the processes by which rock is worn away and ground to sand, by which the sediments that carpet the sea floor are accumulated, shifted, sorted, and distributed. And at death their hard skeletons contribute calcium for the needs of other animals or for the building of the reefs.

Out on the reefs the long-spined black sea urchin excavates cavities along the base of the coral wall; each sinks into its depression and turns its spines outward, so that a swimmer moving along the reef sees forests of black quills. This urchin also wanders in over the reef flats, where it nestles close to the base of a loggerhead sponge, or sometimes, apparently finding no need of concealment, rests in open, sand-floored areas.

A full-grown black urchin may have a body or test nearly 4 inches in diameter, with spines 12 to 15 inches long. This is one of the comparatively few shore animals that are poisonous to the touch, and the effect of contact with one of the slender, hollow spines is said to be like that of a hornet sting, or may even be more serious for a child or an especially susceptible adult. Apparently the mucous coating of the spines bears the irritant or poison.

This urchin is extraordinary in the degree of its awareness of the surroundings. A hand extended over it will cause all the spines to swivel about on their mountings, pointing menacingly at the intruding object. If the hand is moved from side to side the spines swing about, following it. According to Professor Norman Millott of the University College of the West Indies, nerve receptors scattered widely over the body receive the message conveyed by a change in the intensity of light, responding most sharply to suddenly decreased light as a shadowy portent of danger. To this extent, then, the urchin may actually "see" moving objects passing nearby.

Linked in some mysterious way with one of the great rhythms of nature, this sea urchin spawns at the time of the full moon. The eggs and sperm are shed into the water once in each lunar month during the summer season, on the nights of strongest moonlight. Whatever the stimulus to which all the individuals of the species respond, it assures that prodigal and simultaneous release of reproductive cells that nature often demands for the perpetuation of a species.

Off some of the Keys, in shallow water, lives the so-called slate-pencil urchin, named for its short stout spines. This is an urchin of solitary habit, single individuals sheltering under or among the reef rocks near the low-tide level. It seems a sluggish creature of dull perceptions, unaware of the presence of an intruder, and making no effort to cling by means of its tube feet when it is picked up. It belongs to the only family of modern echinoderms that also existed in Paleozoic time; the recent members of the group show little change from the form of ancestors that lived hundreds of millions of years ago.

Another urchin with short and slender spines and color variations ranging from deep violet to green, rose, or white, sometimes occurs abundantly on sandy bottoms carpeted with turtle grass, camouflaging itself with bits of grass and shell and coral fragments held in its tube feet. Like many other urchins, it performs a geologic function. Nibbling away at shells and coral rock with its white teeth, it chips off fragments that are then passed through the grinding mill of its digestive tract; these organic fragments, trimmed, ground, and polished within the urchins, contribute to the sands of tropical beaches.

And the tribes of the starfish and the brittle stars are everywhere represented on these coral flats. The great sea star, Oreaster, stout and powerful of body, perhaps lives more abundantly a little offshore, where whole constellations of them gather on the white sand. But solitary specimens wander inshore, seeking especially the grassy areas.

A small reddish-brown starfish, Linkia, has the strange habit of breaking off an arm, which then grows a cluster of four new arms that are temporarily in a "comet" form. Sometimes the animal breaks across the central disc; regeneration may result in six- or seven-rayed animals. These divisions seem to be a method of reproduction practiced by the young, for adult animals cease to fragment and produce eggs.

About the bases of gorgonians, under and inside of sponges, under movable rocks and down in little, eroded caverns in the coral rock live the brittle stars. With their long and flexible arms, each composed of a series of "vertebrae" shaped like hourglasses, they are capable of sinuous and graceful motion. Sometimes they stand on the tips of two arms and sway in the motion of the water currents, bending the other arms in movements as graceful as those of a ballet dancer. They creep over the substratum by throwing two of their arms forward and pulling up the body or disc and the remaining arms. The brittle stars feed on minute mollusks and worms and other small animals. In turn, they are eaten by many fish and other predators, and sometimes fall victims to certain parasites. A small green alga may live in the skin of the brittle star; there it dissolves the calcareous plates, so that the arms may break apart. Or a curious little degenerate copepod may live as a parasite within the gonads, destroying them and rendering the animal sterile.

My first meeting with a live West Indian basket star was something I shall never forget. I was wading off Ohio Key in water little more than knee deep when I found it among some seaweeds, gently drifting on the tide. Its upper surface was the color of a young fawn, with lighter shades beneath. The searching, exploring, testing branchlets at the tips of the arms reminded me of the delicate tendrils by which a growing vine seeks out places to which it may attach itself. For many minutes I stood beside it, lost to all but its extraordinary and somehow fragile beauty. I had no wish to "collect" it; to disturb such a being would have seemed a desecration. Finally the rising tide and the need to visit other parts of the flat before they became too deeply flooded drove me on, and when I returned the basket star had disappeared.

The basket starfish or basket fish is related to the brittle stars and serpent stars but displays remarkable differences of structure: each of the five arms diverges into branching V's, which branch again, and then again and again until a maze of curling tendrils forms the periphery of the animal. Indulging their taste for the dramatic, early naturalists named the basket stars for those monsters of Greek mythology, the Gorgons, who wore snakes in place of hair and whose hideous aspect was supposed to turn men to stone; so the family comprising these bizarre echinoderms is known as the Gorgonocephalidae. To some imaginations their appearance may be "snaky-locked," but the effect is one of beauty, grace, and elegance.

All the way from the Arctic to the West Indies basket stars of one species or another live in coastal waters, and many go down to lightless sea bottoms nearly a mile beneath the surface. They may walk about over the ocean floor, moving delicately on the tips of their arms. As Alexander Agassiz long ago described it, the animal stands "as it were on tiptoe, so that the ramifications of the arms form a kind of trellis-work all around it, reaching to the ground, while the disk forms a roof." Or again they may cling to gorgonians or other fixed sea growths and reach out into the water. The branching arms serve as a finemeshed net to ensnare small sea creatures. On some grounds the basket stars are not only abundant but associate in herds of many individuals as though for a common purpose. Then the arms of neighboring animals become entwined in a continuous living net to capture all the small fry of the sea who venture, or are helplessly carried, within reach of the millions of grasping tendrils.

To see a basket starfish close inshore is one of those rare happenings that lives always in memory, but it is far otherwise with certain other members of the spiny-skinned tribe of echinoderms—

the holothurians, or sea cucumbers. I have never waded far out onto the flats without meeting them. Their large dark forms, shaped much like the vegetable whose name they have been given, stand out clearly against the white sand where they lie sluggishly, sometimes partly buried. The holothurians perform a function in the sea that is roughly comparable to that of earthworms on land, ingesting quantities of sand and mud and passing it through their bodies. Most of them use a crown of blunt tentacles operated by strong muscles to shovel the bottom sediments into their mouths, then extract food particles from this detritus as it passes through their bodies. Perhaps some calcareous materials are dissolved out by the chemistry of the holothurian body.

Because of their abundance and the nature of their activities, the sea cucumbers profoundly influence the distribution of the bottom deposits around the coral reefs and islands. In a single year, it has been estimated, the holothurians in an area less than two miles square may redistribute 1000 tons of bottom substance. And there is evidence also concerning their work on sea bottoms lying at abyssal depths. The carpeting sediments, which accumulate slowly but unceasingly, lie in orderly layers from which geologists can read many chapters of the past history of the earth. But sometimes the layers are curiously disturbed. Bits of volcanic ash shard originating, for example, from some ancient eruptions of Vesuvius, may in some places lie, not in a thin layer representing and dating the eruption, but widely scattered through the overlying layers of other sediments. Geologists regard this as the work of deep-sea holothurians. And other evidence from deep dredgings and bottom samplings suggests the existence of herds of holothurians on the sea floor at great depths, working over a bottom area, then moving or in a vast migration directed, not by seasonal change, but by the scarcity of food in those deep and lightless regions.

Except in those parts of the world where they are sought as human food (they are the "trepang," or beche-de-mer, of Oriental markets) the sea cucumbers have few known enemies, yet they possess a strange defense mechanism that they employ when strongly disturbed. Then the holothurian may contract strongly and hurl out the greater part of its internal organs through a rupture in the body wall. Sometimes this action is suicidal, but often the creature continues to live and grows a new set of organs.

Dr. Ross Nigrelli and his associates of the New York Zoological Society have recently discovered that the large West Indian sea cucumber (also found about the Florida Keys) produces one of the most powerful of all known animal poisons, presumably as a chemical means of defense. Laboratory experiments showed that even small doses of the poison affect all kinds of animals, from protozoa to mammals. Fish confined in a tank with the cucumber always die when the act of evisceration occurs. The study of this natural toxin reveals the hazardous existence of many small creatures that live in association with another. The sea cucumber attracts a number of such animal associates or commensals. This particular species very often has a small pearl fish, Fierasfer, living within the shelter of the cloacal cavity, which the respiratory activities of the cucumber keep supplied with well-oxygenated water. But the well-being, and indeed the very life of the small Fierasfer seem to be constantly endangered, for the commensal fish is actually living beside a vat of deadly poison that may at any moment be ruptured. Apparently the fish has not developed an immunity to the poison of the holothurian, for Dr. Nigrelli found that if the cucumber was disturbed, its tenant Fierasfer would drift out in a moribund condition, even if actual evisceration did not take place.

Dark patches like the shadows of clouds are scattered over the inshore shallows of the reef flats. Each is a dense growth of sea grass pushing up flat blades through the sand, forming a drowned island of shelter and security for many animals. About the Keys these grass patches consist largely of stands of turtle grass, with which manatee grass and shoal grass may be intermingled. All belong to the highest group of plants—the seed plants—and so are different from the algae or seaweeds. The algae are the earth's oldest plants, and they have always belonged to the sea or the fresh waters. But the seed plants originated on land only within the past 60 million years or so and those now living in the sea are descended from ancestors who returned to it from the land—how or why it is hard to say. Now they live where the salt sea covers them and rises above them. They open their flowers under the water; their pollen is water-borne; their seeds mature and fall and are carried away by the tide. Thrusting down their roots into the sand and the shifting coral debris, the sea grasses achieve a firmer attachment than the rootless algae do; where they grow thickly they help to secure the offshore sands against the currents, as on land the dune grasses hold the dry sands against the winds.

In the islands of turtle grass many animals find food and shelter. The giant starfish, Oreaster, lives here. So do the large pink or queen conch, the fighting conch, the tulip band shell, the helmet shells and the cask shells. A strange, armor-encased fish, the cowfish, swims just above the bottom, parting grass blades to which pipefish and sea horses cling. Baby octopuses hide among the roots and when pursued dive down deep into the yielding sand and disappear from view. Down in that grass-root under-turf many other small beings, of diverse kinds, live deep within the shadowed coolness, to come out only when night and darkness hide them.

But by day many of the bolder inhabitants may be seen by one who wades to the grassy patches and peers down through the clarifying glass of a water telescope, or, swimming above the deeper patches, looks down through a face mask. Here one is most apt to find, in life, the large mollusks that are familiar because their dead and empty shells are common on the beach or in shell collections.

Here in the grass is the queen conch, which in earlier days had a place on almost every Victorian mantel or hearth, and even today is displayed by the hundred at every Florida roadside stand selling tourist souvenirs. Through excessive fishing, however, it is becoming rare in the Florida Keys and is now exported from the Bahamas for use in cutting cameos. The weight and massiveness of its shell, the sharp spire and the heavily armored whorls are eloquent of the defenses raised, through the slow interaction of biology and environment, by myriad ancestral generations. Despite the cumbrous shell and the massive body that thrusts itself out to move over the bottom by grotesque leaps and tumblings, the queen conch seems an alert and sentient creature. Perhaps this effect is heightened by the eyes borne on the tips of two long tubular tentacles. The way the eyes are moved and directed leaves little doubt that they receive impressions of the animal's surroundings and transmit them to the nerve centers that serve in place of a brain.

Although its strength and awareness seem to fit the queen conch for a predatory life, it is probably a scavenger that only occasionally turns to live prey. Its enemies must be comparatively few and ineffectual, but the conch has formed one very curious association. A small fish habitually lives within its mantle cavity. There can be little free space when all of the body and foot are drawn into the shell, but somehow there is room for the cardinal fish, an inch-long creature. Whenever danger threatens, it darts into the fleshy cavern deep within the shell of the conch. There it is temporarily imprisoned when the snail pulls back into its shell and closes the sickle-shaped operculum.

To other, smaller beings that find their way into the interior of the shell, the conch reacts less tolerantly. Current-borne eggs of many sea creatures, larvae of marine worms, minute shrimp or even fish, or non-living particles like grains of sand, may swim or drift inside and, lodging on shell or mantle, set up an irritation. To this the conch responds with ancient defenses, acting to wall off the particle so that it can no longer irritate delicate tissues. The glands of the mantle secrete about this nucleus of foreign matter layer after layer of mother-of-pearl—the same lustrous substance that lines the inside of the shell. In this way the conch creates the pink pearls sometimes found within it.

The human swimmer drifting idly above the turtle grass—if he is patient enough and observant enough—may see something of other lives being lived above the coral sand, from which the thin flat blades of grass reach upward and sway to the motion of the water, leaning shoreward on a flooding tide and seaward on the ebb. If, for example, he looks very carefully he may see what he had thought to be a blade of grass (so perfectly did it simulate one by form and color and movement) detach itself from the sand and go swimming through the water. The pipefish—an incredibly long, slender, and bony-ringed creature that seems quite unfishlike—swims between the grasses slowly and with deliberate movement, now with its body held vertically, now leaning horizontally into the water. The slim head with its long, bony snout is thrust with probing motions into clusters of turtle grass leaves or down among the roots, as the fish searches for small food animals. Suddenly there is a quick, inflating motion of the cheek, and a tiny crustacean is sucked in through the tube-like beak, as one would suck a soda through a straw.

The pipefish begins life in a strange manner, being developed, nurtured, and reared beyond the stage of helpless infancy by the male parent, who keeps his young within a protective pouch. During the mating act of the pair, the ova are fertilized and are placed in this pouch by the female; there they develop and hatch, and to this marsupium the young may return again and again in moments of danger, even long after they are able to swim out into the sea at will.

So effective is the camouflage of another inhabitant of the grass—the sea horse—that only the sharpest eye can detect one at rest, its flexible tail gripping a blade of grass and its bony little body leaning out into the currents like a piece of vegetation. The sea horse is completely encased in an armor composed of interlocking bony plates; these take the place of ordinary scales and seem to be a sort of evolutionary harking back to the time when fish depended on heavy armor to protect them from their enemies. The edges of the plates, where they join and interlock, are produced into ridges, knobs, and spines to form the characteristic surface pattern.

Sea horses often live in vegetation that is floating rather than rooted; such individuals may then become part of that steady northward drift bearing plants, associated animals, and the larvae of countless sea forms into the open Atlantic and eastward toward Europe, or into the Sargasso Sea. Such sea-horse voyagers in the Gulf Stream sometimes are carried ashore on the southern Atlantic coast along with bits of wind- and current-borne sargassum weed to which they cling.

In some of the turtle-grass jungles all of the smaller inhabitants seem to borrow a protective color from their surroundings. I have dragged a small dredge in such a place and found, entangled in the handfuls of grass that came up, dozens of small animals of different species, all an amazing, bright green hue. There were green spider crabs with extremely long, jointed legs. There were small shrimp, also grass-green. Perhaps the most fantastic touch was contributed by several baby cowfish. Like their elders, whose remains one often finds in the debris of the high-tide line, these little cowfish were encased in bony boxes that held head and body in an inflexible case, from which fins and tail protruded as the only movable parts. From tip of tail to the little forward-projecting, bovine horns, these small cowfish were the green of the grass in which they lived.

Especially where they border the channels between the Keys, the shoals carpeted with marine grasses are visited from time to time by sea turtles, which live in some numbers about the outer reef. The hawksbill wanders far out to sea, and seldom turns landward; but the green and loggerhead turtles often swim into the shallow waters of Hawk Channel or seek the passages between the Keys, where the tides race swiftly. When these turtles visit the grassy shoals they are usually seeking those inflated sand dollars, the sea biscuits, whose home is among the grass, or they may seize some of the conchs. Apart from others of their own kind, the conchs probably have no more dangerous enemies than the big turtles.

However far they may wander, loggerhead, green, or hawksbill all must return to land for the spawning season. There are no spawning places on the Keys of coral rock or limestone, but on some of the sand keys of the Tortugas group the loggerhead and the green turtles emerge from the ocean and lumber over the sand like prehistoric beasts to dig their nests and bury their eggs. The chief spawning places of the turtles, however, are on the beaches of Cape Sable and other sand strands of Florida, and farther north in Georgia and the Carolinas.

If the predatory visits of the big turtles to the sea-grass meadows are sporadic, it is far otherwise with the ceaseless, day-by-day preying of the various conchs, one upon another, and all upon mussels or oysters, sea urchins and sand dollars. The chief predator of all the conchs is the dusky-red spindle-shaped one called the horse conch. One has only to see it feeding to realize how powerful it is; when the massive body, brick-red like the shell, is extended to enfold and overwhelm its prey, it seems impossible to believe that so much flesh can ever be drawn back into the shell again. Even the king crown conch, itself a predator on many other conchs, is no match for it. No other American gastropods approach its size. (One-foot individuals are fairly common and the giants of its kind are two feet long.) The big cask shells also are victims of the horse conch, while they themselves feed usually on urchins. Yet I have felt little awareness of this relentless predation on making a casual visit to the habitat of the conchs. There are long periods of somnolence and repletion, and the grassy world by day seems a peaceful place. A conch gliding over the coral sand, a sea cucumber burrowing sluggishly among the roots of the grasses, or the dark and swiftly fleeting forms of sea hares in sudden passage may be the only visible signs of life and motion. For by day life is in retreat; life is buried and hidden in crevices and corners of ledge and rock; life has crept under or within the shelter of sponge or gorgonian or coral or empty shell. In the shallow waters of the shore, many creatures must avoid the penetrating sunlight that irritates sensitive tissues and reveals prey to predator.

But that which seems quiescent—a dream world inhabited by creatures that move sluggishly or not at all—comes swiftly to life when the day ends. When I have lingered on the reef flat until dusk fell, a strange new world, full of tensions and alarms, has replaced the peaceful languor of the day. For then hunter and hunted are abroad. The spiny lobster steals out from under the sheltering bulk of a big sponge and flashes away across the open water. The gray snapper and the barracuda patrol the channels between the Keys and dart into the shallows in swift pursuit. Crabs emerge from hidden caverns; sea snails of varied shape and size creep out from under rocks. In sudden movements, swirling waters, and half-seen shadows that dart across my path as I wade shoreward, I sense the ancient drama of the strong against the weak.

Or if I have listened from the deck of a boat anchored at night among the Keys, I have heard splashings of large bodies moving in the shallows nearby, or the slap of a broad form striking water as a sting ray leaps into the air and falls, leaps and falls again. One of those whom the night stirs to activity is the needle-fish, long, slender, and powerful of body, armed with a sharp beak that would seem more appropriate in a bird. By day the small needle-fish may be seen from wharves and sea walls as they come close inshore, floating at the surface like straws adrift in the water. At night the large fish, that have ranged far to sea, come in to feed in the shallows, sometimes singly, sometimes in large schools. They leap out of the water or go skipping along the surface, making a disturbance that can be heard for a long distance on a calm night. Fishermen say that the needle-fish jump toward a light—that if one is out in a small boat at night where the needle-fish are hunting, it is dangerous, if not suicidal, to show a light, for the fish will leap across the boat. Probably there is an element of truth in the belief, for in some places in the Keys the beam of a searchlight thrown out across the water on a calm night—even if no fish have been heard about—will often be greeted by a series of splashes as a dozen or more large fish leap out of the water. The leaps, however, are usually at right angles to the beam, and the fish seem to be trying to escape the light.

This coral coast is the drowned world of the offshore reef and the world of the shallow reef flats with their fringing, rocky rim; it is also the green world of the mangrove, silent, mysterious, always changing—eloquent of a life force strong enough to alter the visible face of its world. As the corals dominate the seaward margin of the keys, the mangroves possess the sheltered or bay shores, completely covering many of the smaller keys, pushing out into the water to lessen the spaces between the islands, building an island where once there was only a shoal, creating land where once there was sea.

Mangroves are among the far migrants of the plant kingdom, forever sending their young stages off to establish pioneer colonies a score, a hundred, or a thousand miles from the parent stock. The same species live on the tropical coasts of America and the west coast of Africa. Probably the American mangroves crossed from Africa eons ago, via the Equatorial Current—and probably such migrants continue to arrive unnoticed from time to time. How the mangroves got to the Pacific coast of tropical America is an interesting problem. There is no continuous system of currents that would have carried them around the Horn, and besides the cold water to the south would be a barrier. It is not certain how early the mangroves arose, but definite fossil records seem to go back only to the Cenozoic, whereas the Panama Ridge, separating Atlantic from Pacific waters, probably arose much earlier, toward the end of the Mesozoic. By some means, however, the mangroves made the journey to Pacific shores, where they became established. Their further migrations also are mysterious. They must have dispatched their migrant seedlings into the great currents of the Pacific, for at least one American species grows on the islands of Fiji and Tonga and seems to have drifted as well to Cocos-Keeling and Christmas Islands. And some appeared as new colonists on the devastated island of Krakatoa, after it was virtually destroyed by volcanic eruption in 1883.

The mangroves belong to the highest group of plants, the spermatophytes or seed-bearers, whose earliest forms developed on land, and as such they are a botanical example of that return toward the sea that is always fascinating to observe. Among mammals, the seals and whales made such a return to the habitat of their ancestors. The marine grasses have gone even farther than the mangrove, for they live permanently submerged. But why this return to salt water? Perhaps the mangroves or their ancestral stock were forced out of more crowded habitats by the competition of other species. Whatever the reason, they have invaded and established themselves in the difficult world of the shore with such success that no plant now threatens their dominance there.

The saga of an individual mangrove begins when the long pendent green seedling, produced on the parent tree, drops to the floor of the swamp. Perhaps this happens at low tide when all the water has drained away; then the seedling lies amid the tangled roots, waiting till the salt flood comes in to lift it and later float it seaward on the turn of the tide. Of all the hundreds of thousands of red mangrove seedlings produced annually on the southern Florida coast, probably less than half remain to develop near the parent trees. The rest put out to sea, their buoyant structure keeping them in the surface waters, moving with the flow of the currents. They may drift for many months, being able to survive the normal vicissitudes of such a journey—sun, rain, the battering of a rough sea. At first they float horizontally, but with increasing age and the development of their tissues for a new phase of life they gradually come to lie almost vertically with the future root end downward, ready for that contact with earth upon which their future existence depends.

Perhaps in the path of such a pelagic seedling there may lie a small shoal, a little ridge off an island shore, deposited, grain by grain, by the waves. As the tide floats the young mangrove into the shallows, the downward-pressing tip touches the shoal; the sharp point, pressing earthward, becomes embedded. The water movements of later tides rising and falling press the young plant firmly into the receptive soil. Later, perhaps, they bring other seedlings to lodge beside it.

No sooner have the young mangroves anchored themselves than they begin to grow, sending out tiers of roots that arch out and downward to form a circle of supporting props. Among this rapidly increasing tangle of roots, debris of all sorts comes to rest—decaying vegetation, driftwood, shells, coral fragments, uprooted sponges and other sea growths. From such simple beginnings, an island is born.

In twenty to thirty years the young mangroves have acquired the stature of trees. These mature mangroves can resist the battering of a considerable surf, and probably are destroyed only by violent hurricanes. Once in many years such a hurricane comes. Because of the efficiency of their buttressing roots, few mangroves are uprooted even by a violent blow. But the high storm tides press far inland through the swamp, carrying the salt of the open sea into the forested interior. Leaves and small branches are stripped off and carried away, and if the wind is truly violent the trunks and limbs of the great trees are shaken and battered until the bark separates and blows away in sheets, exposing the naked trunk to the burning salt breath of the storm. This may be the history of some of the mangrove ghost forests bordering the Florida coast. But such catastrophes are rare, and in southwestern Florida whole islands of mangroves come to maturity without any serious interruption of their growth.

A mangrove forest, its fringing trees literally standing in salt water, extending back into darkening swamps of its own creation, is full of the mysterious beauty of massive and contorted trunks, of tangled roots, and of dark green foliage spreading an almost unbroken canopy. The forest with its associated swamp forms a curious world. On their flood the tides rise over the roots of the outermost trees and penetrate into the swamplands, carrying many small migrants— the pelagic larvae of sea creatures. Over the ages many of these have found a suitable climate for their survival and have become established, some on the roots or trunks of the mangroves, some in the soft mud of the intertidal zone, some on the bottom of the bay offshore. The mangrove may be the only kind of tree, or the only seed plant growing there; all the associated plants and animals are bound to it by biological ties.

Within the range of the tides the prop roots of the mangroves are thickly overgrown with an oyster whose shell has fingerlike projections to grasp these firm supports and so to remain above the mud. On the night ebb tides, raccoons follow the water down, leaving meandering tracks across the mud as they move from root to root, finding food within the shells of the oysters. The crown conch also preys heavily on these oysters of the mangroves. Fiddler crabs dig tunnels in the mud, sheltering deep within them when the salt tide rises. These crabs are remarkable for the possession, by the males, of one immense claw— the "fiddle"—that is incessantly waved about, apparently serving for communication as well as for defense. Fiddlers eat plant debris picked from the surface of sand or mud. For this the female has two spoon claws; the male, because of his fiddle, only one. By their activities the crabs help to aerate the heavy mud, which is saturated with organic debris and so deficient in oxygen that the mangroves must breathe through their aerial roots to supplement what their buried roots can obtain. Brittle stars and strange burrowing crustaceans live among the roots, while overhead in the upper branches great colonies of pelicans and herons find roosting and nesting places.

Here on these mangrove-fringed shores some of the pioneering mollusks and crustaceans are learning to live out of the sea from which they recently came. Among the mangroves and in marshy areas where the tides rise over the roots of sea grasses there is a small snail whose race is moving landward. This is the coffee-bean shell, a small creature within a short, widely ovate shell tinted with the greens and browns of its environment. When the tide rises the snails clamber up on the mangrove roots or climb the stems of the grasses, deferring as long as possible the moment of contact with the sea. Among the crabs, too, land forms are evolving. The purple-clawed hermit inhabits the strip above the highest tidal flotsam, where land vegetation fringes the shore, but in the breeding season it moves down toward the sea. Then hundreds of them lurk under logs and bits of driftwood, waiting for the moment when the eggs, carried by the female under her body, shall be ready to hatch. At that time the crabs dash into the sea, liberating the young into the ancestral waters. Nearing the end of its evolutionary journey is the large white crab of the Bahamas and southern Florida. It is a land dweller and an air-breather, and it seems to have cut its ties with the sea—all its ties, that is, but one. For in the spring the white crabs engage in a lemming-like march to the sea, entering it to release their young. In time the crabs of a new generation, having completed their embryonic life in the sea, emerge from the water and seek the land home of their parents.

For hundreds of miles this world of swamp and forest created by the mangroves extends northward, sweeping from the Keys around the southern tip of the Florida mainland, reaching from Cape Sable north along the coast of the Gulf of Mexico through all the Ten Thousand Islands. This is one of the great mangrove swamps of the world, a wilderness untamed and almost unvisited by man. Flying above it, one can see the mangroves at work. From the air the Ten Thousand Islands show a significant shape and structure. Geologists describe them as looking like a school of fish swimming in a southeasterly direction— each fish-shaped island having an "eye" of water in its enlarged end, the heads of all the little "fish" pointing to the southeast. Before these islands came to be, one may suppose, the wavelets of a shallow sea heaped the sand of its floor into little ridges. Then came the colonizing mangroves, converting the ripple marks to islands, perpetuating in living green forest the shape and trend of the sand ripples.

Today we can see, from one generation of man to another, where several small islands have coalesced to form one, or where the land has grown out and an island has merged with it—sea becoming land almost before our eyes.

What is the future of this mangrove coast? If it is written in its recent past we can foretell it: the building of a vast land area where today there is water with scattered islands. But we who live today can only wonder; a rising sea could write a different history.

Meanwhile the mangroves press on, spreading their silent forests mile upon mile under tropical skies, sending down their grasping roots, dropping their migrant seedlings one by one, launching them into the drifting tides on far voyages.

And offshore, under the surface waters where the moonlight falls in broken, argent beams, under the tidal currents streaming shoreward in the still night, the pulse of life surges on the reef. As all the billions of the coral animals draw from the sea the necessities of their existence, by swift metabolism converting the tissues of copepods and snail larvae and minuscule worms into the substance of their own bodies, so the corals grow and reproduce and bud, each of the tiny creatures adding its own limy chamber to the structure of the reef.

And as the years pass, and the centuries merge into the unbroken stream of time, these architects of coral reef and mangrove swamp build toward a shadowy future. But neither the corals nor the mangroves, but the sea itself will determine when that which they build will belong to the land, or when it will be reclaimed for the sea.

VI
THE ENDURING SEA

NOW I HEAR the sea sounds about me; the night high tide is rising, swirling with a confused rush of waters against the rocks below my study window. Fog has come into the bay from the open sea, and it lies over water and over the land's edge, seeping back into the spruces and stealing softly among the juniper and the bayberry. The restive waters, the cold wet breath of the fog, are of a world in which man is an uneasy trespasser; he punctuates the night with the complaining groan and grunt of a foghorn, sensing the power and menace of the sea.

Hearing the rising tide, I think how it is pressing also against other shores I know—rising on a southern beach where there is no fog, but a moon edging all the waves with silver and touching the wet sands with lambent sheen, and on a still more distant shore sending its streaming currents against the moonlit pinnacles and the dark caves of the coral rock.

Then in my thoughts these shores, so different in their nature and in the inhabitants they support, are made one by the unifying touch of the sea. For the differences I sense in this particular instant of time that is mine are but the differences of a moment, determined by our place in the stream of time and in the long rhythms of the sea. Once this rocky coast beneath me was a plain of sand; then the sea rose and found a new shore line. And again in some shadowy future the surf will have ground these rocks to sand and will have returned the coast to its earlier state. And so in my mind's eye these coastal forms merge and blend in a shifting, kaleidoscopic pattern in which there is no finality, no ultimate and fixed reality—earth becoming fluid as the sea itself.

On all these shores there are echoes of past and future: of the flow of time, obliterating yet containing all that has gone before; of the sea's eternal rhythms—the tides, the beat of surf, the pressing rivers of the currents—shaping, changing, dominating; of the stream of life, flowing as inexorably as any ocean current, from past to unknown future. For as the shore configuration changes in the flow of time, the pattern of life changes, never static, never quite the same from year to year. Whenever the sea builds a new coast, waves of living creatures surge against it, seeking a foothold, establishing their colonies. And so we come to perceive life as a force as tangible as any of the physical realities of the sea, a force strong and purposeful, as incapable of being crushed or diverted from its ends as the rising tide.

Contemplating the teeming life of the shore, we have an uneasy sense of the communication of some universal truth that lies just beyond our grasp. What is the message signaled by the hordes of diatoms, flashing their microscopic lights in the night sea? What truth is expressed by the legions of the barnacles, whitening the rocks with their habitations, each small creature within finding the necessities of its existence in the sweep of the surf? And what is the meaning of so tiny a being as the transparent wisp of protoplasm that is a sea lace, existing for some reason inscrutable to us—a reason that demands its presence by the trillion amid the rocks and weeds of the shore? The meaning haunts and ever eludes us, and in its very pursuit we approach the ultimate mystery of Life itself.

|APPENDIX: CLASSIFICATION|

Protophyta, Protozoa: One-celled Plants and Animals

THE SIMPLEST FORMS of cellular life are the one-celled plants (Protophyta) and one-celled animals (Protozoa). In both groups, however, there are many forms that defy attempts to place them definitely in one category or another because they display characteristics usually considered animal-like along with others usually thought definitive of plants. The Dinoflagellata form such an indeterminate group, and are claimed both by zoologists and by botanists. Although a few are large enough to be seen without magnification, most are smaller. Some wear shells with spines and elaborate markings. Some have a remarkable, eye-like sense organ. All dinoflagellates are immensely important in the economy of the sea as food for certain fishes and other animals. Noctiluca is a relatively large dinoflagellate of coastal waters, where it produces brilliant displays of phosphorescence, or by day reddens the water Sphaerella the abundance of its pigmented cells. Other species are the cause of the phenomenon known as "red tide," in which the sea is discolored and fishes and other animals die from poisons given off by the minute cells. The red or green scum of high tide pools, "red rain," and "red snow" are growths of these forms, or of green algae (e.g., Sphaerella). Much phosphorescence or "burning" of the sea is caused by dinoflagellates, which create a uniformly diffused light, lacking large spots of illumination. Examined closely, in a vessel of water, the light is seen to consist of tiny sparks.

The Radiolaria are one-celled animals whose protoplasm is contained in siliceous shells of extraordinary beauty. These minute shells, sinking to the bottom, accumulate there to form one of the characteristic oozes or sediments of the sea floor. The Foraminifera are another unicellular group. Most have calcareous shells, though some build their protective structures with sand grains or sponge spicules. The shells, eventually drifting to the floor of the ocean, cover vast areas with calcareous sediments that, through geologic change, may become compacted into limestone or chalk, and raised to form such features of the present landscape as the chalk cliffs of England. Most Foraminifera are so minute that one gram of sand might contain up to 50,000 shells. On the other hand, a fossil species, Nummulites, was sometimes 6 or 7 inches across and formed limestone beds in Northern Africa, Europe, and Asia. This limestone was used in the building of the Sphinx and the great pyramids. Fossil Foraminifera are much used by geologists in the oil industry in correlating rock strata.

Diatoms (Greek, diatomos—cut in two) are minute plants usually classified among the yellow-green algae because they contain granules of yellow pigment. They exist as single cells or in chains of cells. The living tissue of a diatom is encased within a shell of silica, of which one half fits over the other, as a lid over a box. Fine etchings on the surface of the shell create beautiful patterns and are characteristic for the various species. Most diatoms live in the open sea, and because they exist in inconceivable abundance are the most important single food stuff in the ocean, being eaten not only by many small animals of the plankton, but by many larger creatures, as mussels and oysters. The hard shells sink to the bottom after the death of the tissues, and accumulate there to form diatom oozes that cover vast areas of ocean floor.

The blue-green algae, or Cyanophyceae, are among the simplest and oldest forms of life and are the most ancient plants that still exist. They are widely distributed and occur even in hot springs and other places where conditions are so difficult that no other plant life can exist. They often multiply in phenomenal numbers, giving the surface of ponds and other still waters a colored film known as water bloom. Most are encased in gelatinous sheaths that protect them from extreme heat or cold. They are well represented in the "black zone" above high-tide line on rocky shores.

Thallophyta: Higher Algae

THE GREEN ALGAE, or Chlorophyceae, are able to endure strong light and thrive high in the intertidal zone. They include such familiar forms as the leafy sea lettuce and a stringy, tube-like alga of high rocks and tide pools called Enteromorpha ("intestine-shaped"). In the tropics some of the most common green algae are the brush-shaped Penicillus that forms minute groves over the coral reef flats, and the beautiful little cup alga, Acetabularia, like tiny, inverted mushrooms of purest green. Some of the green algae of the tropics are important in the economy of the sea as concentrators of calcium. Although the group is most typical of warm, tropical seas, the green algae are found on the shore wherever there is strong sunlight, and others of the group live in fresh water.

The brown algae, or Phaeophyceae, possess various pigments that conceal their chlorophyll, so their prevailing colors are brown, yellowish, or olive-green. They are largely absent from warmer latitudes except in deep water, being unable to endure heat and strong sun. An exception is the Sargassum weed of tropical shores, which drifts northward in the Gulf Stream. On northern coasts the brown rockweeds live between tide lines, and the kelps or oarweeds from the low-tide line down to depths of 40 to 50 feet. Although all of the algae select and concentrate in their tissues many different chemicals present in sea water, the brown seaweeds and especially the kelps are extraordinary in the quantity of iodine stored. Formerly they were utilized widely in the industrial production of iodine. The same seaweeds now are important in the production of the carbohydrate algin for use in fire-resistant textiles, jellies, ice cream, cosmetics, and various industrial processes. The presence of alginic acid gives these seaweeds their great resilience in heavy surf.

The red algae, or Rhodophyceae, most sensitive light of all the seaweeds, send only a few hardy species (including Irish moss and dulse) into the intertidal zone; most are delicate and graceful seaweeds living for the most part below low water. Some live deeper than any other seaweeds, going down into the dim regions 200 fathoms or more below the surface. Some (the corallines) form hard crusts on rocks or shells. Containing magnesium carbonate as well as calcium carbonate, these algae seem to have played an important geochemical role in earth history, perhaps having aided the formation of the magnesium-rich marble dolomite.

Porifera: Sponges

THE SPONGES (Porifera, or pore-bearers) are among the simplest of animals, being little more than an aggregation of cells. Yet they have gone a step beyond the Protozoa, for there are inner and outer layers of cells, with some hint of specialization of function—some for drawing in water, some for taking in food, some for reproduction. All these cells cohere and work together to carry out the single purpose of the sponge—to pass the waters of the sea through the sieves of its own being. A sponge is an elaborate system of canals contained in a matrix of fibrous or mineral substance, the whole pierced by numerous small entrance pores and larger exit holes. The inmost or central cavities are lined with flagellated cells that remind one of protozoan flagellates. The lashing of the whiplike flagella creates currents to draw in water. In passage through the sponge, the water gives up food, minerals, and oxygen, and carries away waste products.

To a certain extent, each of the smaller groups within the sponge phylum has a physical appearance and habit of life that is characteristic, yet the sponges are probably more plastic in relation to their environment than any other animals. In surf they take the form of a flattened crust, almost without regard to species; in deep, quiet water they may assume an upright tubular form, or branch in a way suggestive of shrubbery. Their shape, therefore, is little or no aid in identification, and the classification of sponges is based chiefly on the nature of their skeleton, which is a loose network of minute hard structures called spicules. In some the spicules are calcareous. In others they are siliceous, although sea water contains only a trace of silica and the sponge must have to filter prodigious quantities to obtain enough for its spicules. The function of extracting silica from sea water is confined to primitive forms of life, and among animals does not occur above the sponges. Commercial sponges fall into a third group, having a skeleton of horny fibers. They are confined to tropical waters.

From such a beginning toward specialization, nature seems to have gone back and made a fresh start with, other materials. All evidence points toward a separate origin for the coelenterates and all other more complex animals, leaving the sponges in an evolutionary blind alley.

Coelenterata: Anemones, Corals, Jellyfish, Hydroids

THE COELENTERATES, despite their simplicity, foreshadow the basic plan on which, with elaborations, all the more highly developed animals are formed. They possess two distinct layers of cells, the outer ectoderm and the inner endoderm, sometimes with an undifferentiated middle layer that is not cellular but is the forerunner of the third cell layer, the mesoderm, of the higher groups. Each coelenterate is basically a hollow double-walled tube, closed at one end and open at the other. Variations of this plan have resulted in such diverse forms as the sea anemones, corals, jellyfish, and hydroids.

All coelenterates possess stinging cells called nematocysts, each of which is a coiled, pointed thread contained in a sac of turgid fluid, ready to be expelled to impale or entangle passing prey. Stinging cells are not developed in higher animals; although they have been reported in flatworms and sea slugs, they have been secondarily acquired by eating coelenterates.

The Hydrozoa display most clearly another peculiarity of this group, known as alternation of generations. An attached, plantlike generation produces a medusoid generation, shaped like small jellyfish. These, in turn, produce another plantlike generation. In the hydroids the more conspicuous generation is an attached, branching colony bearing tentacled individuals, or hydranths, on its "stems." Most of these are shaped like small sea anemones and capture food. Other individuals bud off the new generation—tiny medusae that (in many forms) swim away, mature, and shed eggs or sperm cells into the sea. An egg produced by such a medusa, when fertilized, develops into another plantlike stage.

In another group, the Scyphozoa, or true jellyfish, the plantlike generation is the inconspicuous one, and the medusae are highly developed. The jellyfish range from very small creatures to the immense arctic jelly, Cyanea, which reaches an extreme diameter of 8 feet (1 to 3 feet is more common) with tentacles up to 75 feet long.

In the Anthozoa (flower animals) the medusoid generation has been completely lost. This group includes the anemones, corals, sea fans, and sea whips. The anemone represents the basic plan; all the rest of this group are colonial forms in which the individual, anemone-like polyps are embedded in some sort of matrix, which may be stony, as in the reef-building corals, or, in the sea fans and sea whips, may consist of a horny substance of protein nature, similar to the keratin of vertebrate hair, nails, and scales.

Ctenophora: Comb Jellies

THE ENGLISH WRITER Barbellion once said that a comb jelly in sunlight is the most beautiful thing in the world. Its tissues are almost crystal clear, and as this little ovoid creature twirls in the water it flashes iridescent lights. The ctenophores, or comb jellies, are sometimes mistaken for jellyfish because of their transparency, but there are various structural differences, with the "comb-plates" being characteristic of the phylum. These occur in eight rows on the outer surface. Each plate has a hinged attachment and bears hairlike cilia along its free edge; as the plates flash in succession to propel the animal through the water, the cilia break up the rays of sunlight and produce the characteristic flashing.

Like some of the jellyfish, most ctenophores possess long tentacles. These are equipped not with stinging cells, but with sticky pads that capture prey by entanglement. Ctenophores eat enormous numbers of fish fry and other small animals. They live chiefly in the surface waters.

The ctenophores comprise a small phylum, with less than 100 species. Members of one of their groups have flattened bodies and do not swim, but creep on the ocean floor. Some specialists believe these creeping ctenophores have given rise to the flatworms.

Platyhelminthes: Flatworms

THE FLATWORMS include many parasitic as well as many free-living forms. Leafy thin, the free-living flatworms flow like a living film over rocks or sometimes swim by flapping undulations in a way reminiscent of skates. They have made significant advances in an evolutionary sense. They are the first to possess three primary layers of cells, a characteristic of all higher animals. They also have a bilateral type of symmetry (one side being a mirror image of the other), with a head end that always goes first. They have the simple beginnings of a nervous system and eyes that may be only simple pigment spots or, in some species, well-developed organs with lenses. There is no circulatory system, and perhaps it is because of this that all flatworms have such thin bodies, in which all parts are in easy communication with the exterior, and oxygen and carbon dioxide are easily passed through surface membranes to underlying tissues.

Flatworms are found among seaweeds, on rocks, in tide pools, and lurking in dead mollusk shells. They are usually carnivorous, devouring worms, crustaceans, and mollusks of minute size.

Nemertea: Ribbon Worms

THE RIBBON WORMS have extraordinarily elastic bodies, sometimes round, sometimes flat. One of them, the bootlace worm (Lineus longissimus) of British waters, may attain a length of 90 feet and is the longest of all the invertebrates. The American Cerebratulus of shallow coastal waters often is 20 feet long and about an inch wide. Most, however, are only a few inches long and many are considerably less than an inch. They habitually contract into coils or knots when disturbed.

All ribbon worms are highly muscular but lack the co-ordination of nerve and muscle that higher worms have. There is a brain consisting of simple nerve ganglia. Some have primitive hearing organs, and the characteristic slits along the sides of the head (suggestive of a mouth) seem to contain important organs of sensation. Although there are a few hermaphroditic species, in most ribbon worms the sexes are separate. There is, however, a strong tendency toward asexual reproduction, and associated with this is a habit of breaking up into many pieces when handled. The fragments then regenerate complete worms. Professor Wesley Coe of Yale University found that a certain species of ribbon worm could be cut repeatedly until eventually miniature worms less than one one-hundred thousandth the volume of the original were obtained. An adult can live a year without food, according to Professor Coe, compensating for lack of nourishment by diminishing in size.

The ribbon worms are unique in the possession of an extensible weapon called a proboscis, enclosed in a sheath and capable of being suddenly everted, hurled out, and coiled around the prey, which is then drawn back toward the mouth. In many species the proboscis is armed with a sharp lance, or stylet, which if lost is quickly replaced by another held in reserve. All ribbon worms are carnivorous, and many prey on the bristle worms.

Annelida: Bristle Worms

THE ANNELID (ringed, or segmented) worms include several classes, one of which, the Polychaeta (many bristles) includes most marine annelids. Many of the polychaetes, or bristle worms, are active swimmers that make their living as predators; others are more or less sedentary, building tubes of various sorts in which they live, either feeding on detritus in sand or mud or on plankton which they strain from the water. Some of these worms are among the most beautiful creatures of the sea, their bodies shining with iridescent splendor, or adorned with feathery crowns of tentacles in soft and beautiful colors.

In their structure they represent a great advance over lower forms. Most of them possess a circulatory system (although the blood worm, Glycera, much used as bait, has no blood vessels but a blood-filled cavity between the skin and the alimentary canal) and so are able to dispense with the thinness of body of the flatworms, for the blood flowing through vessels transports food and oxygen to all parts of the body. The blood is red in some, green in others. The body consists of a series of segments, several of the anterior ones being fused to form the head. Each segment bears a pair of unbranched, unsegmented paddle-like appendages for crawling or swimming.

Bristle worms include many diverse forms. The familiar nereids, or clam worms, often used for bait, spend most of their lives in crude burrows among stones on the sea bottom but emerge to hunt or, in swarms, to spawn. The sluggish scale worms live under rocks, in muddy burrows, or among the holdfasts of seaweeds. The serpulid worms build variously shaped limy tubes from which only their heads emerge; other worms, like the beautifully plumed Amphitrite, form mucous tubes under rocks or crusts of coralline algae or on muddy bottoms, and a worm of colonial habit, Sabellaria, uses coarse sand grains to build elaborate structures that may be several feet across. Though honeycombed with the burrows of the worms, these massive dwelling places are strong enough to bear the weight of a man.

Arthropoda: Lobsters, Barnacles, Amphipods

THE ARTHROPOD (jointed foot) phylum is an enormous group, comprising five times as many species as are included in all the rest of the animal phyla combined. The arthropods include the crustacea (e.g. crabs, shrimps, lobsters), the insects, the myriopods (centipedes and millipedes), the arachnids (spiders, mites, and king crabs) and the tropical, wormlike Ony-chophora. All marine arthropods belong to the class Copepod Crustacea except for a scant handful of insects, a few mites and sea spiders, and the king crabs.

Whereas the paired appendages of the annelids are simple flaps, those of the arthropods possess multiple joints and are specialized to perform such varied functions as swimming, walking, handling food, and gaining sensory impressions of the environment. Whereas the annelids interpose only a simple cuticle between their internal organs and the environment, the arthropods protect themselves by a rigid skeleton of chitin impregnated with lime salts. This, in addition to being protective, has the advantage of giving a firm support for the insertion of muscles. On the other hand there is the disadvantage that, as the animal grows, the rigid outer covering must be shed from time to time.

The crustaceans include such familiar animals as crabs, lobsters, shrimps, and barnacles, as well as less-known creatures like the ostracods, isopods, amphipods, and copepods, all of which are important or interesting for one reason or another.

The ostracods are unusual arthropods in that they are not segmented but are enclosed in a two-part carapace, or shell, flattened from side to side, and opened and closed by muscles like a mollusk's shell. The antennae act as oars and are extended through the opened carapace to row the little animal through the water. Ostracods often live in seaweeds or in sand on the ocean floor, usually being quiet by day and coming out to feed at night. Many marine ostracods are luminous and as they swim about emit little puffs of bluish light. They are one of the chief sources of phosphorescence at sea. Even when dead and dried they retain the phosphorescent quality to an astonishing degree. Professor E. Newton Harvey of Princeton University says in his authoritative volume Bioluminescence that during the Second World War Japanese army officers used dried ostracod powder in advanced positions where use of flashlights was prohibited—by adding a few drops of water to a little powder in the palm of the hand, they could obtain enough light to read dispatches.

Copepods (oar-footed) are very small crustaceans with rounded bodies, jointed tails, and oarlike legs with which to propel themselves jerkily. In spite of their minute size (from microscopic to half-inch) the copepods form one of the basic populations of the sea, and are food for an immense variety of other animals. They are an indispensable link in the food chain by which the nutrient salts of the sea are eventually made available (via plant plankton, animal plankton, carnivores) to larger animals such as fishes and whales. Copepods of the genus Calanus, known as "red feed," redden large areas of ocean surface and are eaten in prodigious numbers by herring and mackerel and also by certain whales. Birds of the open sea such as petrels and albatrosses are plankton feeders and sometimes subsist largely on copepods. In their turn, the copepods graze on diatoms, eating sometimes as much as their own weight in a day.

Amphipods are small crustaceans that are flattened from side to side, while isopods are flattened from upper to lower surface. The names are a scientific reference to the kinds of appendages possessed by these small creatures. The amphipods have feet that can be used both for swimming and walking or crawling. The isopods, or "equal-footed" animals, have appendages that show little difference in size and shape from one end of the body to the other.

On the shore the amphipods include the beach hoppers, or sand fleas, that rise in clouds (leaping, not flying) from masses of seaweed when they are disturbed, and others that live offshore in seaweed and under rocks. They eat fragments and bits of organic debris and are themselves eaten in great number by fish, birds, and other larger creatures. Many amphipods wriggle along on their sides when out of water. Sand hoppers use their tails and posterior legs as a spring and progress by leaps; other species swim.

Isopods of the shore (closely related to the familiar sow bugs of the garden) include the slaters (sea roaches, wharf rats, quay lice) often seen running over rocks and wharf pilings. These have left the water and seldom return to it; they drown if long submerged. Others live offshore, often in seaweeds whose color and form they mimic. Still others swarm in tide pools, sometimes nipping the skin of human waders to produce a tingling or itching sensation.

Most are scavengers; some are parasites; and some form habitual associations (commensalism) with an animal of unrelated species.

Both amphipods and isopods carry their young in brood chambers instead of liberating eggs into the sea. This habit has helped some in each group to live high on the shore and is a necessary preliminary to land existence.

The barnacles belong to the order Cirripedia (Latin, cirrus—a ringlet or curl), presumably named because of their gracefully curving feathery appendages. The larval stages are free-living and resemble the larvae of many other crustaceans, but the adults are attached, living in a shell of calcareous material, fixed to rocks or other hard objects. The gooseneck barnacles are attached by a leathery stalk; the rock or acorn barnacles are attached directly. The gooseneck barnacles are often oceanic, attaching themselves to ships and floating objects of all sorts. Some of the acorn barnacles grow-on the hide of whales or the shells of sea turtles.

The large crustaceans—shrimps, crabs, and lobsters—not only are most familiar but display the typical arthropod plan of body most clearly. The head and thoracic regions usually are fused and covered with a hard shell, or carapace; only the appendages indicate the division into segments. The flexible abdomen or "tail," on the other hand, is divided into segments and usually is an important aid to swimming. Crabs, however, keep the tail segments folded under the body. The hard shell of an arthropod must be shed periodically as the animal grows. The creature gets out of the old shell through a slit that opens up usually across the back. Underneath is the new shell, much folded and wrinkled, soft and tender. The crustacean, after shedding, may spend days in seclusion, hiding from enemies until its armor has hardened.

The class Arachnoidea includes in one group the horseshoe crabs, and in another diverse one the spiders and mites, only a few of which are marine. The horseshoe, or king, crab has a peculiar distribution, being very abundant on the Atlantic coast of America, absent from Europe, and represented by three species on the Asiatic coast from India to Japan. Its larval stages closely resemble the ancient trilobites of Cambrian times and as a reminder of those past ages it is often called a living fossil. Horseshoe crabs are abundant along the shores of bays and other relatively quiet waters, where they eat clams, worms, and other small animals. They come out on beaches early in the summer to lay eggs in depressions scooped out in the sand.

Bryozoa: Moss Animals, Sea Laces

THE BRYOZOA are a group of uncertain position and relationships, including rather diverse forms. They may appear as fluffy plantlike growths often mistaken for seaweeds, especially when found dried on the shore. Another form grows as flat hard patches encrusting seaweeds or rocks and having a lacy appearance. Still another type is a branched and upright growth of gelatinous texture. All of these are colonial forms or associations of many individual polyps, all living in adjoining cells or embedded in a unifying matrix.

The encrusting Bryozoa, or sea laces, are beautiful mosaics of closely set compartments, each inhabited by a small tentacled creature that superficially resembles the hydroid polyp, but possesses a complete digestive system, a body cavity, simple nervous system, and many other features of higher animals. The individuals of a bryozoan colony are largely independent of each other, instead of being connected as the hydroids are.

The Bryozoa are an ancient group dating from the Cambrian. They were considered seaweeds by early zoologists, and later were classified as hydroids. There are about 3000 marine species, compared with only about 35 in fresh waters.

Echinodermata: Starfish, Sea Urchins, Brittle Stars, Sea Cucumbers

OF ALL the invertebrates, the echinoderms are most truly marine, for among their nearly 5000 species not one lives in fresh water or on land. They are an ancient group, dating from the Cambrian, but in all the hundreds of millions of years since then none has even attempted to make the transition to a land existence.

The earliest echinoderms were the crinoids, or sea lilies, stalked forms that lived attached to the floor of Paleozoic seas. Some 2100 fossil species of crinoids are known, in contrast to about 800 living species. Today most crinoids live in East Indian waters; a few occur in the West Indian region and come as far north as Cape Hatteras, but there are none in the shallow waters of New England.

The common echinoderms of the shore represent the four remaining classes of the phylum: the sea stars, the brittle and serpent stars, the sea urchins and sand dollars, and the holothurians, or sea cucumbers. In all members of the group there is a recurrent insistence on the number five, many of the structures occurring in fives or multiples of five, so that the figure is almost a symbol of the group.

The sea stars, or starfish, have flattened bodies, many in the conventional five-pointed shape, though the number of arms varies. The skin is roughened by hard limy plates from which short spines grow. In most species the skin also bears structures like minute forceps on flexible stalks (called pedicellaria); with these the animal keeps the skin clear of sand grains and also picks off larvae of sedentary forms that try to settle there. This is necessary because the delicate breathing organs—soft rosettes of tissue—also project through the skin.

Like all other echinoderms, the starfish possess a so-called water-vascular system that functions in locomotion and secondarily in other ways, and consists of a series of water-filled tubes running to all parts of the body. Intake of sea water is accomplished, in starfish, through a conspicuous perforated plate on the upper surface—the madreporite (mother of pores). The fluid passes along the water canals and eventually into the many short flexible tubes (tube feet) that occupy the long grooves on the under surface of the arms. Each tube bears a sucker at its tip. The tube feet can be lengthened or contracted by changes of hydrostatic pressure—when extended, the suckers grip the underlying rock or other hard surface and the animal pulls itself along. The tube feet are used also to grip the shells of mussels or other bivalve mollusks on which the starfish preys. As the starfish moves, any of its various arms may Brittle star go first and thus serve as temporary "head."

In the slender, graceful brittle stars and serpent stars the arms are not grooved and the tube feet are reduced. However, these animals progress rapidly by writhing motions of the arms. They are active predators and feed on a variety of small animals. Sometimes they lie in "beds" of many hundreds of animals on the sea bottom offshore—a living net through which scarcely any small creature can safely reach bottom.

In the sea urchins the tube feet are arranged in five avenues or rows running from upper to lower apexes of the body, just as the meridians on a globe run from pole to pole. The skeletal plates of the urchins are articulated rigidly to form a globular shell, or test. The only movable structures are the tube feet, which are thrust out through perforations in the test, the pedicellaria, and the spines, which are mounted on protuberances on the plates. The tube feet are retracted when the animal is out of water, but when submerged they may be extended beyond the spines to grasp the substratum or to capture prey. They may also perform some sensory functions. In the various species the spines differ greatly in length and thickness.

The mouth is on the under surface, surrounded by five white, shining teeth used to scrape vegetation off the rocks and also to assist in locomotion. (Although other invertebrates—e.g., the annelids—have biting jaws, the urchins are the first to have grinding or chewing organs.) The teeth are operated by an internally projecting apparatus of calcareous rods and muscles known to zoologists as Aristotle's lantern. On the upper surface the digestive tract opens to the exterior through a centrally placed anal pore. Around this are five petal-shaped plates, each bearing a pore that serves to discharge eggs or sperm. The reproductive organs are arranged in five clusters just under the upper or dorsal surface. They are practically the only soft parts the animal possesses and it is for these that the sea urchins are sought as human food, especially in Mediterranean countries. Gulls hunt the urchins for a similar purpose, often dropping them on the rocks to break the tests so that they can eat out the soft parts.

The eggs of the sea urchins have been used extensively in biological studies of the nature of the cell, and Jacques Loeb in 1899 used them in a historic demonstration of artificial parthenogenesis, causing an unfertilized egg to develop merely by treating it with chemicals or by mechanical stimulus.

The holothurians, or sea cucumbers, are curious echinoderms with soft, elongated bodies. They crawl on one surface with the mouth end foremost and so have secondarily substituted a functional bilateral symmetry for the radial symmetry characteristic of the phylum. Tube feet, where present, are confined to three rows on the functional under surface of the body. Some holothurians are burrowing forms, using small spicules embedded in the body surface to grasp the surrounding mud or sand and aid their progress. The shapes of these spicules vary with the species and often must be studied microscopically before correct identification can be made. The holothurians are large and abundant in tropical seas (they are the trepang, or beche-de-mer, of commerce) and in northern waters are represented by smaller species living on offshore bottoms or among intertidal rocks and seaweeds.

Mollusca: Clams, Snails, Squids, Chitons

BECAUSE OF their endlessly varied shells, often intricately made and beautifully adorned, some of the mollusks probably are better known than any other animals of the shore. As a group they possess qualities different from those of any other invertebrates, although their more primitive members and the nature of their larvae suggest that their remote ancestors may have resembled those of the flatworms. They have soft, unsegmented bodies typically protected by a hard shell. One of the most remarkable and characteristic molluscan structures is the mantle, a cloaklike tissue that encloses the body, secretes the shell, and is responsible for its complex structure and adornment.

The most familiar mollusks are the snail-like gastropods and the clamlike bivalves. The most primitive mollusks are the creeping, sluggish coat-of-mail shells, or chitons, the least known are the tusk shells, or scaphopods, and the most highly developed class the cephalopods, represented by the squids.

The shells of the gastropods are univalve or in one piece, and coiled in more or less spiral fashion. Nearly all snails are "right-handed," that is, the opening is to the right as it faces the observer. One of the exceptions is the "left-handed conch," one of the most common gastropods of Florida beaches. Occasionally a left-handed individual occurs in a normally right-handed species. Some gastropods have reduced the shell to an internal remnant, as in the sea hares, or have lost it entirely, as in the sea slugs or nudibranchs (in which, however, a coiled shell is present in the embryo).

The snails are for the most part active animals, both the vegetarians that move about scraping plant food from the rocks and the carnivores that capture and devour animal prey. The sedentary boat shells, or slipper shells, are exceptions; they attach themselves to shells or to the sea bottom and live on diatoms strained from the water, in the manner of oysters, clams, and other bivalves.

Most snails glide about on a flattened muscular "foot," or they may use this same organ to burrow into the sand. When disturbed, or at low tide, they draw back into their shells, the opening being closed by a calcareous or horny plate called the operculum. The shape and structure of the operculum vary greatly in the different species and sometimes it is useful in identification. In common with other mollusks (except the bivalves) the gastropods have a remarkable, tooth-studded band, the radula, on the floor of the pharynx, or, in some species, on the end of a long proboscis. The radula is used to scrape off vegetation or to drill holes in shelled prey.

The bivalves, with few exceptions, are sedentary. Some (e.g., the oyster) fix themselves permanently to a hard surface. Mussels and some others anchor themselves by secreting silklike byssus threads. The scallops and the lima clams are examples of the few bivalves that possess the ability to swim. The razor clams have a slender pointed foot by means of which they dig deeply and with incredible speed into the sand or mud.

Bivalves that bury deeply in the substratum are able to do so because they possess a long breathing tube, or siphon, through which they draw in water and so receive oxygen and food. Although most are suspension feeders, filtering minute food organisms from the water, some, including the tellins and coquina clams, live on detritus that accumulates on the sea floor. There are no carnivorous bivalves.

The shells of gastropods and bivalves are secreted by the mantle. The basic chemical material of molluscan shells is calcium carbonate, which forms the outer layer of calcite, and the inner layer of aragonite, which is a heavier and harder substance although it has the same chemical composition. Calcium phosphate and magnesium carbonate also are contained in mollusk shells. The limy materials are laid down on an organic matrix of conchiolin, a substance chemically allied to chitin. The mantle contains pigment-forming cells as well as shell-secreting cells. The rhythm of activity of these two kinds of cells results in the marvelous sculpturing and color patterns of molluscan shells. Although shell formation is affected by many factors in the environment and in the physiology of the animal itself, the basic hereditary pattern is so strongly determined that each species of mollusk has its characteristic shell by which it may be identified.

A third class of the mollusk phylum consists of the cephalopods, so unlike the snails and clams that superficially it is hard to reconcile the relationship. Although ancient seas were dominated by shelled cephalopods, all but one (the chambered nautilus) have now lost the external shell, retaining only an inconspicuous internal remnant. One large group, the decapods, have cylindrical bodies with ten arms; they are represented by the squids, the ramshorn shell, and the cuttlefish. Another group, the octopods, have baglike bodies with eight arms; examples are the octopus and the argonaut.

The squids are strong and agile; over short distances they are probably the swiftest animals of the sea. They swim by expelling a jet of water through the siphon, controlling the direction of motion by pointing the siphon forward or backward. Some of the smaller species swim in schools. All squids are carnivorous, preying on fish, crustaceans, and various small invertebrates. They are sought by cod, mackerel, and other large fish, and are a favorite bait. The giant squid is the largest of all invertebrates. The record specimen, taken on the Grand Banks of Newfoundland, measured about 55 feet including the arms.

Octopuses are nocturnal animals and, according to those most familiar with their habits, are timid and retiring. They live in holes or among rocks, feeding on crabs, mollusks, and small fish. Sometimes the location of an octopus den may be discovered by the pile of empty mollusk shells near the entrance.

The chitons belong to a primitive order of mollusks, the Amphineura. Most of them wear a shell consisting of eight transverse plates bounded by a tough band, or girdle. They creep sluggishly over rocks, scraping off vegetation. At rest, they settle into a depression, blending so well with their surroundings that they are easily overlooked. They are sought as food (sea beef) by West Indian natives.

The fifth class of mollusks consists of the little-known scaphopods (tooth shells or tusk shells), which form shells resembling an elephant's tusk, from one to several inches long and open at both ends. They dig into sandy bottoms, using a small, pointed foot. Some specialists think their structure may be similar to that of the ancestors of all mollusks. However, this is a field for speculation, since the principal classes of mollusks were all defined early in the Cambrian, and clues to the nature of the ancestral forms are exceedingly vague. The tooth shells number about 200 species, and are widely distributed in all seas. None, however, are intertidal.

Chordata: Subphylum Tunicata

THE ASCIDIANS, or sea squirts, are the most common representatives on the shore of that interesting group of early chordates, the Tunicata. As forerunners of the vertebrates, or backboned animals, all of the chordates have at some time a stiffening rod of cartilaginous material, an evolutionary forecast of the vertebral column which all the higher animals were to possess. The adult ascidian paradoxically suggests a creature of low and simple organization, with a physiology somewhat like that of oysters or clams. It is only in the larva that the chordate characteristics are clear. Though minute, the larva strongly resembles the tadpole of a frog, possessing a notochord and a tail and swimming actively. At the end of the larval period it settles down, becomes attached, and undergoes metamorphosis to the much simpler adult form, in which the chordate characters are lost. This is a curious phenomenon of evolution, which seems to be degenerative rather than progressive, with the larva displaying more advanced characteristics than the adult.

The adult sea squirt is shaped like a bag with two tubular openings or siphons for water intake and outgo, and a pharynx perforated with many slits through which water is strained. The common name refers to the fact that when the animal is disturbed it contracts sharply, forcing jets of water out through the siphons. In the so-called simple ascidians the animals live as separate individuals, each enclosed in a tough covering or test of material chemically akin to cellulose. Sand and debris often adhere to this test, forming a mat in which the actual shape of the animal is seldom apparent. In this form they often grow profusely on wharf pilings, floats, and rocky ledges. In the compound, or colonial, type of ascidian many individuals live together, embedded in a tough gelatinous substance. Unlike a group of simple ascidians, the various individuals of a colony are derived by asexual budding from one individual, the founder of the colony. One of the commonest compound sea squirts is the sea pork, Amaroucium, named from the usually gray, gristly appearance of its colonies. These may form a thin mat on the under side of a rock or, offshore, grow erect, forming thick slabs that may break off and be carried in to shore. The individuals composing the colony are not easily seen, but under a lens pits in the surface appear, each the opening through which a single sea squirt communicates with the outside world. In the beautiful compound sea squirt Botryllus, however, the individuals form flowerlike clusters, easily visible.